The Anatomist

The ANATOMIST

A TRUE STORY OF *Gray's Anatomy*

Bill Hayes

Hypo-glossal N.

Crico-thyroid artery.

BALLANTINE BOOKS

NEW YORK

the neck, showing the carotid and subclavian arteries.

Published in the United States by Ballantine Books, an imprint of The Random House Publishing Group, a division of Random House, Inc., New York.

BALLANTINE and colophon are registered trademarks of Random House, Inc.

Grateful acknowledgment is made to St. George's Hospital Library, London, for permission to reproduce the photographs of Henry Gray on pages xix and 229; to the Wellcome Library, London, for permission to reproduce the illustrations on pages 71, 102, 214–215, and 231; to the President and Council of the Royal College of Surgeons of England for permission to reproduce the illustration on page 121 (from MS. 0134, Student Diary of Henry Vandyke Carter); to The British Library, London, for permission to reproduce the photographs on pages 188 and 189, © The British Library Board; and to Martin Duke for permission to reproduce the photograph of Henry Gray's tombstone on page 229. All other images are in the public domain or are courtesy of the author.

LIBRARY OF CONGRESS CATALOGING-IN-PUBLICATION DATA
Hayes, Bill.
The anatomist : a true story of Gray's anatomy / Bill Hayes.
p. ; cm.
Includes bibliographical references and index.
ISBN: 978-0-345-45689-2 (hardcover : alk. paper)
1. Gray, Henry, 1825–1861. 2. Gray, Henry, 1825–1861. Anatomy.
3. Carter, H. V. (Henry Vandyke), 1831–1897. 4. Anatomists—England—Biography. I. Title.
[DNLM: 1. Gray, Henry, 1825–1861. 2. Carter, H. V. (Henry Vandyke), 1831–1897. 3. Anatomy—Biography. 4. Anatomy—Personal Narratives. 5. Anatomy—History. 6. History, 19th Century. 7. Students—Biography. 8. Students—Personal Narratives. QS 11.1 H417a 2008]
QM16.G73H39 2008
611.0092—dc22   2007028733

Printed in the United States of America on acid-free paper

www.ballantinebooks.com

9 8 7 6 5 4 3 2

FIRST EDITION

Book design by Simon M. Sullivan

For Steve Byrne

Contents

Prologue

LOOKING BACK, I CAN SEE HOW MY WHOLE LIFE HAS LED TO THIS: a book about a book about anatomy and about the education of an anatomist, albeit an amateur one. Sigmund Freud was right, it turns out: *Anatomy is destiny*—or mine, at least.

A bloom in a boulder crack, my fascination with human anatomy took root in the less-than-accommodating conditions of a strict Irish Catholic upbringing in the 1960s. *You are made in God's image,* I remember being told by the nuns in catechism classes, which struck me as wonderful news; to cherish your body was to cherish the Creator. At the same time, though, the story of Adam and Eve made it frighteningly clear that the body is a shameful vessel for sin. Even today, while I no longer consider myself Catholic or even religious, the tale of their fall from innocence haunts me: God warning Adam, you shall die if you eat of the tree of knowledge, and then Eve—poor, gullible Eve—sweet-talked by the snake, pulling an apple from the tree. I still just want to stop her—*No!*

"Then the eyes of both were opened; and they knew that they were naked; and they sewed fig leaves together." Banishment from the garden was but one part of their sentence. "You are dust," God tells them, "and to dust you shall return."

The moral, simple enough for a child to grasp, is that when God says no, he really means no. But the story also conveys a more insidious notion: awareness of the body may lead to spiritual death.

To the eight-year-old me, fresh from making my first confession, Adam and Eve were especially effective in promoting the idea that nakedness went hand in hand with sin. And yet, making matters morally confusing, there were some naked people it was okay to look at, whose nakedness you were meant to take notice of, beginning, ironically enough, with Adam and Eve. Even in my children's Bible, those two appeared as delectable as a couple of ripe Red Delicious apples. The most frequent naked body I saw while growing up, though, belonged to Jesus. In our house, crucifixes were as common as light fixtures. A small bronze one hung above my bed, and I prayed to it every night. But, in a curious design choice, as I think of it now, the largest crucifix was posted right outside the bathroom my five sisters and I shared. Jesus, as if clad in a towel rather than a loincloth, appeared to be waiting his turn for the shower. I can still recall every detail of that crucifix, a wooden one my dad had bought in Mexico. The body was carved with such care, so that legs and arms were finely muscled and veined and the torso made long and sinuous. His nakedness exposed every crucifixion wound and was crucial to reinforcing a central tenet of the church: The gash along his ribs was due to our sin. The trickle of blood down his forehead was our fault. Christ's pain was meant to cause you the same. His death, we were never to forget, was for us.

Providing a ballast to the Irish Catholicism of my father was my mother. Mom had once been an aspiring painter in New York City before meeting Dad, and she was not Catholic. Only on the rarest occasions did Mom join us at church. I remember how, every year on Ash Wednesday, the first day of Lent, when a thumb-press of ash was placed on your forehead as a reminder of your mortality, Mom's unsmudged brow marked her as unlike my father and sisters and me. Dad would jokingly call her "a heathen" but, almost in the same breath, say earnestly to his six children, "Mom's a saint—that's why she doesn't need to go to Mass."

To me, Mom represented the everyday, but also another, higher world—a world of artists; of passionate, driven people; a world I glimpsed in her little library of art books. Above the table where her sewing machine sat was a pinewood bookshelf that held histories of famous painters as well as exhibition catalogs from far-off places such as the Museum of Modern Art in New York. While the book *Picasso's Picassos* only confused me, the thick tomes on Leonardo da Vinci, Michelangelo, and Matisse introduced me to the sensual body as art. These books were full of nudes, not naked people, a distinction I began to understand as I edged toward puberty.

Lining the shelf on the opposite wall was our 1965 *World Book* encyclopedia, twenty-two volumes, straight-spined and orderly, like the cadets in a photo nearby: the 1949 graduating class of West Point, with Dad standing third from the left, front row. It was in *World Book* volume H that I got my first peek inside the human body. Between entries on *hairstyling* and *hysterectomy*, there was a spectacular anatomical illustration composed of ten bright transparent overlays. The body illustrated was male, although, in a nod to modesty, no genitalia were shown. To this day, I still recall the smell of the plastic sheets and the sticky sound they made when you turned each overlay. Sometimes I would run up to one of my sisters and flash Encyclopedia Man in her face, eliciting a guaranteed *ick!!*—this form of teasing worked especially well on Julia, three years younger than me, and Shannon, two years older—but we would then often sit down and look at the illustrations together, drawn into the illusion of a deep body adventure, as though we wore X-Ray Specs that actually worked. Paging from left to right performed a crude dissection, salmon-colored muscle giving way to the wet worms of viscera giving way to less and less until, finally, on the last transparency, only the unadorned skeleton remained.

My two best friends' dads were both doctors, one a G.P., one a dermatologist. Their family bookshelves held volumes that I would never be able to find even at the Spokane Public Library: old medical textbooks. Kept on topmost shelves, they were meant to be out of reach, out of sight, which is of course exactly why I would urge Chris or Andy to fetch them. What I will never forget is the deformities and disfigurements pictured: photos, as artless as mug shots, of elephantiasis, leprosy, gargantuan tumors, and other conditions that made the body seductively grotesque.

Though I confided this to neither Chris nor Andy nor any of my sisters, I dreamed of becoming a doctor one day. But whether because I did so poorly in high school biology and chemistry or because I did so well in English and writing classes, I eventually shelved the idea of a medical career. Still, my interest in the workings of the body remained; indeed, I think it intensified in direct proportion to my burgeoning interest in sex. But by the time I was actually having sex, after moving to San Francisco in the early 1980s, the body had turned virtually overnight into something to fear, a vessel not for mortal sin but for a deadly virus. That was when I bought my first copy of *Gray's Anatomy*.

I got it for the pictures: hundreds of drawings of lean muscle, bones, and organs, each meticulously rendered and labeled as if it were a rare entomology specimen. Lying on a bookstore table, the thick volume's cover image had first drawn me in: a profile of a man whose face is intact but whose neck is not, to put it mildly. The skin from the chin to the collarbone is missing, revealing strips of muscle and a tangle of blood vessels. As gruesome as it was, I found the image incongruously beautiful. The young man wore such a serene expression, and there was something so intimate in his pose—the way his head was gently turned to expose every detail, as if in invitation: *Here, come closer, take a look.*

Marked down to $9.95, the book was also a deal I could not pass up. *Gray's Anatomy*, like *Bulfinch's Mythology* or Plato's *Republic,* seemed a classic every person should have—if only just to have—so I bought a copy. This was almost exactly half my life ago. Aside from occasionally spell-checking anatomical terms while writing my two previous books, *Sleep Demons* and *Five Quarts,* I ended up using the book most often to ID parts on their way out: A good friend's cancerous pancreas. My sister's uterus, at the time of her partial hysterectomy. My boyfriend's pituitary gland tumor. Being able to picture the affected organs helped put those surgeries into clearer focus. *Gray's Anatomy* became like the list of emergency numbers taped next to the phone—good in a crisis. Whenever I used the book, its language, the opposite of lyrical, always impressed me; no detail seemed too small to be harpooned with a handful of finely honed sentences. Such occasions, though, were few and far between. Like my copy of *Bulfinch's Mythology, Gray's* mainly collected dust on a shelf.

One day a few years ago, however, I pulled it out to double-check a spelling, and as I paged through the text, the word in question slipped from my mental grasp and a new thought surfaced: *Who wrote this thing?*

I found nothing on the jacket flap but his first and last names, Henry Gray. There was no "About the Author" page in the back of the book. Curious, I checked an encyclopedia and other reference sources at home. Still nothing. Surely there must be a biography of the man, I thought, as I logged on to the public library's online catalog. Alas, "No matches found." So, too, said Amazon as well as those usually trusty procurers of the obscure, the International League of Antiquarian Booksellers, which struck me as odd. *Gray's Anatomy* is widely considered one of the most famous books in the English language and is the only medical text most people know by name. *Gray's* has been cited as a major influence by figures ranging from fitness icon Jack LaLanne to the artists Jean-Michel Basquiat and Kiki Smith. Fascinating "biographies" have been written about everything from the number zero to the color mauve, yet there is not one on Gray. Can he simply have gone unnoticed by historians, been taken for granted, as he had by me till now?

Well, no, there was a far more reasonable explanation: when trying to piece together the life of Henry Gray, the unknowns simply outnumber the knowns. What I discovered through additional digging at local libraries, in fact, is that surviving records are more extensive about his father. Thomas Gray, born in Hampshire, England, in 1787, was a private messenger to King George IV, a position in which he was entrusted to carry the most sensitive of documents, including, one can assume, the love letters sent back and forth between George and Maria Anne Fitzherbert, the king's secret, illegal, and Roman Catholic wife. Thomas Gray later served in the same capacity for George's successor, William IV, which suggests that he possessed a remarkable ability to be discreet and inconspicuous—a trait that he seems to have passed on to his son. To this day, it is not known where or when exactly Henry Gray was born. The year 1825 has been suggested, but 1827 is generally more

St. George's Hospital

agreed upon. Similarly, where he spent his earliest years is uncertain. Some historians cite London while others suggest Windsor, England, while others, connecting imaginary dots, say the lad was raised in Windsor Castle, a commoner among royalty, which is an enchanting notion but nevertheless wrong. When Henry was around three years old, Thomas Gray moved his family—wife Ann and Henry and his three siblings—into a house near Buckingham Palace, but the address itself, No. 8 Wilton Street, is the only verifiable fact of his boyhood. It is almost as if Henry Gray did not fully exist as a flesh-and-blood being until the sixth of May 1845, the day he stepped inside London's St. George's Hospital and signed his name to the register as a medical student.

From this point forward, one can retrace his path by way of exams passed, awards won, and scientific papers published, all of which are noted in official records. Among the major milestones: Gray received the equivalent of an M.D. in 1848, and four years later, at the precocious age of twenty-five, he was elected a Fellow of the Royal Society, an exclusive group of scientists that had counted Isaac Newton and Antoni van Leeuwenhoek among its members. Still, I found it maddening that I could not scrape together more. While I had long since missed my chance at a medical career, I'd begun to hope I could make a contribution to the field as Henry Gray's biographer. But what I had gathered about him thus far would amount to little more than a CV. No anecdotes or reminiscences about him seemed to have survived. His rise through the ranks of St. George's is marked merely by the titles and dates of his positions, as if the man, like one of his book's drawings, were just a neatly tagged specimen: postmortem examiner (1848), curator of the Anatomical Museum (1852), lecturer in anatomy (1854), and so forth.

Critically praised author was added to the list in 1858. *Anatomy, Descriptive and Surgical,* as Gray's tome was originally titled, received excellent reviews, sold well, and was picked up by an American publisher the following year, by which point he was already working on a revised and enlarged second edition. In what came as a complete surprise to me, however, Gray did not create any of the book's nearly four hundred signature anatomical drawings. These were the work of another Henry, one Henry Vandyke Carter, whose contribution was not even credited in the 1901 American edition of the book that I own. While this revelation raised a slew of new questions, others were put to rest when I learned how Henry Gray's story ended: On Wednesday, June 12, 1861, he was scheduled to appear before the board of governors of St. George's Hospital. As one of three finalists for a prestigious surgical position, he was expected to make a brief statement on his own behalf. But he never showed up. And word eventually reached the panel as to why. Henry Gray had died that very same day. When all the details emerged, it turned out that he had contracted smallpox while treating his young nephew who was suffering from the disease. Ten-year-old Charles Gray recovered and went on to live into his fifties, but Henry, who had been vaccinated against smallpox in childhood, died at his family's longtime home, one week after falling ill. He was thirty-four years old.

At the time of his death, Gray reportedly had completed a substantial portion of a major new book, though this unfinished manuscript has never turned up. Even the original manuscript and drawings for *Gray's Anatomy* have disappeared (most likely, I learned, they had gone up in flames when a fire decimated the British publisher's archives the year Gray died). I probably would have left it at that—my curiosity about Henry Gray more than satisfied, my dream of contributing to medical history properly deferred—had I not come across one last thing: a photograph included in the one hundredth–anniversary edition of his *Anatomy*.

Taken fifteen months before his death, the photo shows Gray and two dozen young men grouped in what looks like a large art studio, with a high vaulted ceiling and drawings pinned to the walls. Sunlight pours down through the banks of skylights. Some standing, some seated, many of the young men have on long white smocks over their suits and ties—one even sports a beret—yet they wear uniformly solemn expressions, as if bearers of grim diagnoses. None is more serious, though, than Henry Gray. He is seated on a stool in the foreground, next to one of several low tables. A diminutive man with dark, deep-set eyes and thick, wavy hair, he looks like a pint-sized Heathcliff. Brooding and intense, he stares at the camera, waiting the long seconds for the shutter to close. This is, of course, a class photo, and no one holds the pose better than the cadaver lying just to Henry Gray's right. Poking out from under a covering, its pale, narrow feet protrude over the table's edge.

I could not get this picture out of my head: the spacious chamber bathed in daylight; the dead body on the table, its upper half sliced off by the picture's edge; and, most of all, the anatomist himself. Something about the look on Gray's face seized my imagination in a way that I can only compare—odd as this may sound—to love at first sight: an overpowering desire to get to know this man as thoroughly as possible. My course of action then seemed perfectly clear: I would come to know Henry Gray by coming to know human anatomy.

Henry Gray and his anatomy students, St. George's Hospital, 1860
PHOTOGRAPH BY JOSEPH LANGHORN

PART ONE *THE STUDENT*

Self-knowledge can, and ought, to apply not only to the soul,
but also to the body;
the man without insight into the fabric of his body
has no knowledge of himself.

—JOHN MOIR, student of anatomy, notes from opening lecture,
Anatomical Education in a Scottish University, 1620

One

O N THE FIRST DAY OF CLASS, I AM MISTAKEN FOR A TEACHING
assistant six times, which, on the one hand, simply tells me I'm
old—a good twenty years older than the average student—but, on
the other hand, seems to imply that I look as if I belong. Choosing
the glass half full, I smile through each mistaken identity.

The class size is 120 (150 if you count the cadavers). We had been
warned that some students are overwhelmed by the first sight of the
dead bodies. And sure enough, some students clearly are. But I am
more freaked out by the woman in the gas mask. *What does she know
that the rest of us don't?*

"Class? Hello?" comes a disembodied voice, tinnily amplified.
This is Sexton Sutherland, one of the three professors, although I
cannot see him for the crowd. "Before we get started, some housekeeping rules . . ."

The first thing he mentions is the color-coded wastebaskets: red is
for tissue (the human type) and white is for regular garbage, and,
please, please don't mix them up. Likewise with the sinks: use only
the stainless steel for this and the porcelain for that, though I cannot
catch the specifics for all the rustling. The mention of first-aid proto-
col finally brings the room to complete silence. And when Dr.
Sutherland directs everyone's attention to the emergency biohazard
showers in each corner of the lab, I find a sea of eyes sweeping over
me, as I happen to be standing right next to one of them. *Towel, anyone?*

"Finally, just some basic etiquette for the weeks to come: No eating your lunch in here." This elicits a collective *ewwwww.* "No music.
Please don't take any pictures. And try to keep your voices down.

Laughter's okay," Dr. Sutherland adds. "We love laughter in the lab—it's a great way to release emotions. But not at the expense of the wonderful people who've donated their bodies to our program." He lets that sink in for a moment. "Okay, let's get going."

A class orientation had been held the day before in a lecture hall downstairs. Afterward, we were invited to check out the lab and, as Dr. Sutherland had said in a masterful sweep of understatement, "to get comfortable with 'the surroundings,' " by which he meant the reclining dead. About half the class had made the trip up to the thirteenth floor, myself included. I was anxious to put glimpsing the cadavers for the first time behind me. And I am glad I did.

If that was the orientation, however, this is more like disorientation. I am not sure what to do or where to go exactly, so I grab the crisp new scrubs from my gym bag, pull them over my head, and join the large group being led by Dana Rohde, interim director of the anatomy course for the University of California–San Francisco School of Pharmacy, whom I had met earlier. Using one cadaver as a demo model, she gives a brief overview of the afternoon's assignment; pauses to explain how to put a fresh blade onto a scalpel; does a quick scan to see that we are all wearing the mandatory rubber gloves; and adds finally, "I'll be back to see how you're doing in half an hour." Dr. Rohde then stands there for a moment, wearing the look of a swimming instructor who finds her class still standing on the deck of the pool: *Why aren't you wet yet?*

Six of us arrange ourselves around cadaver number 4, but rather than looking at the naked female body lying before us, we all stare at one another.

"I haven't dissected anything since high school biology," one of the three women admits, breaking the ice. "And that was a frog."

This seems like the right moment to make an admission of my own: "I should tell you, I am not a student here. Dr. Rohde gave me permission to come to your lectures and labs. I'm just going to be an observer."

All but one of them look as though they would pay to change places with me. Gergen, the exception, a tall, husky, hairy guy who says he has never dissected anything in his life, cheerfully volunteers to begin the dissection. Now, technically, it will be Gergen's first cut, but not this body's. Like all the cadavers used in this ten-week class in gross anatomy, it was worked on during a previous course. Instead of fresh bodies like those routinely autopsied on *CSI*—blue-lipped and gray but still lifelike—these are closer to something from a Discovery Channel special. The cadavers are shrunken like unwrapped Egyptian mummies. The skin, where still intact, is tan and leathery, and the exposed inner flesh is as dark and dried as beef jerky. The heads, hands, and feet are wrapped in strips of gauze, which gives the impression that they had been badly burned. As Dr. Sutherland explained during the orientation, the gauze serves two functions: it helps preserve the delicate parts for a longer period, and it also protects us, in a sense.

"It's usually most *impactful* to see the hands or the face," he had said, treading carefully with his words, "because that's really what represents a person's identity." When dissecting other parts, one quickly learns to dissociate, but this is much harder when you see the eyes or the mouth. Emotions can come up unexpectedly, he then added. "Sometimes, you'll be dissecting away—maybe you're halfway through the course—and then you'll remove a piece of gauze and there's a tattoo and you just stop cold. Or maybe you see nail polish." Any individualizing mark is a stark reminder that this is not just a body but *some*body. As Dr. Sutherland had explained, this is one reason why the first dissection is in a relatively neutral location, the thorax, otherwise known as the chest.

Though I am the sole spectator here today, I take comfort in knowing I am well represented in history. Human dissection has been a riveting spectacle for centuries, and the curious, whether by invitation or paid ticket, have long pressed into crowded rooms, craning necks and breathing through perfumed handkerchiefs, to witness that first ghastly slice, then the next, and the next. In Europe, the need to create a space conducive to teaching, learning, and observing resulted in the Western world's first "anatomical theater," built in Italy in 1594 at the University of Padua. A steeply raked am-

Theater of Anatomy, London, 1815

phitheater that accommodated three hundred, it became the model for other facilities that sprang up at competing schools, including the College of Physicians in London. Always at the center was the dissecting table, with the first circle of spectators barely a blood spurt removed. At UCSF, I and my fellow novice anatomists stand not in a theater but in a no-frills lab. In order to get the best view of what is being dissected at our table, I have to perch on the rungs of a metal stool.

Our cadaver, who in life probably stood no more than five foot two, does not bear the classic "Y" incision of an autopsy (shoulders to sternum, then straight down the abdomen to the pubis). Instead, a kind of double doorway was incised in her chest: the skin cut across the collarbones as well as beneath the ribs—roughly marking the top and bottom of the thorax—and then sliced down the middle.

Before making a new incision, we need to "unpack" the previous work. As Laura reads instructions from the lab guide, Gergen folds back the two large panels of skin, then grasps the edges of the underlying breastplate, a solid shield of ribs and muscles that had been precut with a surgical saw. Gergen lifts, and a fresh wave of fumes escapes from the cadaver, making all of us flinch.

Peering down, I can see why the thorax was once known as the "pantry" of the body. It is a deep, squarish cavity packed full of various objects, one of which Gergen must now remove: a lung. He slips his left hand into the cavity and feels for "the root of the lung," a short, fat tube that is not at the bottom of the lung, as one might imagine a root should be, but toward the top, connecting it to the windpipe. "Now what?" Gergen asks.

Laura, who is as small and slim as Gergen is large, scrambles to find the next instruction. "Let's see here—'Cut through the root of the lung superiorly and continue inferiorly through the pulmonary ligament.' "

"Translation?"

"Top to bottom—slice it off—I think."

Although Gergen does the actual cutting, the rest of us, in spirit at least, help him hold the scalpel steady: Laura, Amy, Miriam, and Massoud are the fingers folded in around him, and I, opposite them, am the thumb. Gergen then steps back, indicating to Laura that she may do the honors. Biting her lower lip, she reaches into the thoracic cavity and, after a little tugging, frees the right lung. The size of a wadded-up T-shirt, it looks like a wet mound of gray taffeta. All six of us wear identical triumphant smiles, as if we have delivered a baby.

But it turns out our baby is ugly. Dr. Rohde returns and points out that the cut was "too lateral," which means the bronchus (an offshoot of the windpipe) is not clearly exposed. But she immediately tries to reassure us. "The only way you learn is by doing it, by making mistakes. Anyway, there are a lot of bodies here to look at, and, luckily, you're not being graded on your surgical skills."

Before moving on to the next table, Dana instructs us on the next task: resection of a half-foot-long section of the phrenic nerve, a narrow fiber running through the thorax, a portion of which is visible now that the right lung is out of the way. Explaining the nerve's primary function in the living, she breaks it down in simple terms: "If it's damaged, you can't breathe." Likewise, if you sever your spinal cord *above* the level of the phrenic, she adds, you lose all use of this nerve. "That's what happened to the actor Christopher Reeve, which is why he had to spend the rest of his life on a ventilator."

At this moment, everyone at our table is having the same illogical reaction: terror that we might render our dead body a quadriplegic. It is halfway through our first three-hour lab, and none of us feels any detachment whatsoever.

Massoud, taking over from Gergen, does not wear the expression of a lucky man, and yet the opportunity before him—to dissect and, yes, even make mistakes—truly is a privilege. To put this into perspective, Hippocrates, the "Father of Medicine," for instance, never dissected a human body because the practice was forbidden in ancient Greek society. Aristotle, too, never broke this taboo, and, jumping ahead to the second century A.D., neither did the revered Greek physician Galen. Galen, whose writings remained medical gospel for fourteen hundred years after his death, had gained his knowledge of anatomy from dissecting pigs and cats. Brilliant but mistaken, he believed that animal and human anatomy were often interchangeable. And like a dropped figure in a checkbook registry, this error only compounded with time.

Human dissection continued to be forbidden in virtually every society on through the Middle Ages. Not that it was not done, I'd wager—the dead body of a stranger surely must have proved too tempting for some unscrupulous practitioners—but how would you share your findings without implicating yourself? In parts of Europe, even dissection of animals eventually fell into disrepute because of its association with sorcery. In the year 1240, however, a radical change in policy took effect. Frederick II, emperor of the Holy Roman Empire, decreed that, for the sake of public health and the training of better doctors, at least one human body would be dissected in his kingdom every five years. For this bold move, Frederick II is credited with single-handedly pulling the field of anatomy out of the dark ages.

By the beginning of the fourteenth century, human dissections were conducted as often as once a year at the top European universities. The corpses used, male and female alike, were almost always those of executed criminals. The leading anatomist of the time, Mondino dei Liucci (c. 1270–c. 1326), a professor at the University of Bologna, became the Henry Gray of the late Middle Ages. His dissection manual, *Anathomia*, completed in 1316, was used in nearly all medical schools throughout Europe for the following two hundred years. After the invention of printing, Mondino's *Anathomia* went through thirty-nine editions, a number that the British version of *Gray's Anatomy* has only just matched.

Mondino earned a place in medical history by performing the first "properly recorded" dissection of a human corpse, but he is also remembered for sparking a revolution in the teaching of anatomy. Mondino systematized the process of dissection, providing a step-by-step method for exploring the human body. Following his lead, later pioneers would eventually overturn many of the fallacies of Galenism. In a sense, Mondino provided the map, allowing his successors to uncover a string of treasures.

In the Mondino method, a human dissection followed a strict schedule dictated by a grim fact: the process was a race against putrefaction. In an age when cadavers were not embalmed, only the cold could slow decomposition, but only somewhat, so the procedure would be carried out during the coldest time of the year and at a rapid clip, over four successive days. Rather than beginning with the outer chest and progressively moving deeper into the body, as one would today, Mondino always dissected from the inside out, starting with the intestines, since they rotted quickly and smelled worst first. Seated above the cadaver on a pulpit, he would recite from his text while the actual cutting was done by a trained assistant. The students never dissected. A second assistant, called a demonstrator, would hold aloft or point out the body parts described. Inciden-

Illustration from an edition
of *Anathomia* by Mondino dei
Liucci, c. 1493

tally, Henry Gray was a member of a similar three-person team at St. George's and over his tenure filled each of these roles.

By the final day of a Mondino dissection, the smell had probably risen to the level of olfactory bludgeon. For this reason, the University of Bologna made a special allowance to the anatomy department, providing a budget to purchase wine for the students and spectators at dissections—a little something to help deaden the senses, one gathers. (Interestingly, the cadaver, too, might have benefited from the alcohol, which, as anatomists would later discover, makes a pretty good preservative.) One final allowance deserves mention. In what may have been the creepiest way in history to earn extra credit, students at Bologna could bring in bodies of their own. But even in this case, they were not permitted to dissect them.

As I step back and watch Massoud, Laura, and the others finish exposing the phrenic nerve, I find myself preoccupied with how tiny our cadaver looks—smaller than any of the others in the room. For a moment, I even wonder if this could be a child, but I know that's not possible; children's bodies are almost never given to an anatomy program (instead, parents will commonly donate a deceased child's organs for transplant or research purposes). A walk to the "Cause of Death" list posted on the back wall sets me straight. We actually have the body of a frail woman who was eighty-eight years old. She died of heart failure and had also had Alzheimer's disease.

Returning to the dissection table, I take the opportunity to feel her lung, which Laura had placed beside her neck. This is the first internal organ I have ever held in my hands. Whereas I thought the lung would feel hollow and light, instead the tissue is dense, with the consistency of a wet loofah. The base of the lung is smooth and con-cave where it had nested upon the top of the diaphragm. I really want to see what the organ looks like on the inside. But that, I trust, will come with another class.

I fold back into the group as they reassemble the chest cavity and notice something startling: the gauze wrapping has fallen away from the cadaver's right hand. The fingernails, a part of the body ex-tremely slow to decay, are still those of a well-groomed little old lady—nicely rounded and buffed, as if she had just come from a manicure. I lift her wrist and the whole arm rises stiffly. I rewrap her hand in gauze, then help pull the drape over her body.

AFTER CLASS, I cross Parnassus Avenue and move from the realm of rubber gloves to white cotton ones, from the dissection laboratory to the Special Collections Room of UCSF's medical library. I have an appointment with a first edition. Up two flights of stairs, the jewel box of a room is climate-controlled and silent, and, save for the librarian and me, empty. Ms. Wheat retreats to a back room in her familiar way and reappears moments later. I love this almost cere-monial part of my visits, the way she approaches my table with the requested volume in her gloved hands, as if she were a sommelier cradling a rare vintage. With a whispered *thank you,* I nod in ap-proval as she places before me an 1858 copy of Gray's *Anatomy.*

The book rests on a large foam pad, angled like a lectern but deeper near the center to minimize stress on the spine. For a nearly 150-year-old book, it is in amazingly good shape. As I admire the pris-tine brown leather cover, I pull on the thin white gloves Ms. Wheat has left me and can't help noticing how similar they are to the ones my sisters each wore to their First Holy Communion. I crack open the cover and turn the first few pages. This releases the faint earthy smell unique to very old books, a smell I happen to like, a scent pre-served from another time.

Although his book has assumed the mantle of a classic, Henry

Gray wrote it for a most prosaic purpose: to satisfy a pressing need for new medical textbooks. The demand was driven by several factors, but the most compelling was the discovery of anesthesia in its earliest form, chloroform. Nowadays, when "going under the knife" is a phrase that's slipped into casual conversation and surgery is entertainment on reality shows, it is hard to imagine how revolutionary it was to suddenly have the ability to safely put patients under, to be able to cut into their flesh without their feeling that burn of the blade. Prior to this innovation, the field of surgery was chiefly concerned with—as paradoxical as this sounds—*external* medicine, what the doctor could see or easily feel under the skin, whether this was a boil to lance, rotten tooth to pull, or gangrenous limb to remove. Since the patient was conscious, a surgeon had to be dexterous and, above all, speedy. With the use of anesthesia, operating theaters became far quieter, doctors could take more time, and an all-new terrain opened up. As never before, doctors had access to deeper, heretofore unreachable areas of the body. Consequently, the scope of what a medical student had to learn grew exponentially; hence, the need for an exhaustive encyclopedia such as Gray's *Anatomy*.

Of course, anatomy texts had been around for more than five hundred years by this point; Henry Gray was not inventing the wheel here. And in fact, several decent textbooks were already available. Quain's *Elements of Anatomy*, for instance, was in its sixth successful edition at the time. But Gray had clear ideas on how to make a better book, and a commercially successful one. The main selling point would be its emphasis on *surgical anatomy*—applying anatomical knowledge to the practice of surgery. This alone would make Gray's *Anatomy* a great buy—a practical text that would remain useful long after the student entered the professional world.

His author credit forms two lines on the title page, bold and capped:

HENRY GRAY, F.R.S.,
LECTURER ON ANATOMY AT ST. GEORGE'S HOSPITAL

For me, seeing it in its original form is equivalent to being formally introduced to this man whom, till now, I had known only from a distance. The introductions continue in the introduction itself, a section not reproduced in my copy of the book. Here, Gray acknowledges the contributions of two friends: Timothy Holmes, who helped edit the text, and Henry Vandyke Carter, who both executed the drawings and assisted with the many, many dissections required for the work.

I tear a piece of scrap paper to mark this page and nearly give Ms. Wheat a coronary. Her expression is somewhere between cat-in-bathwater and teacher-on-edge. With a pursed expression, she promptly delivers to my left hand a pile of precut page markers.

I had brought with me to the library my copy of *Gray's* (a 1901 facsimile) so I could compare them side by side. What's immediately obvious is that hundreds of drawings by a different artist were added

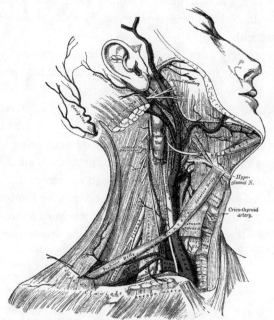

Fig. 383.—Surgical anatomy of the arteries of the neck, showing the carotid and subclavian arteries.

to the later version, although, even without the benefit of credits, it is easy for me to tell whose work was whose. When the book was first published, the British medical journal *The Lancet,* typically not one to rave, praised it as the best anatomy treatise "in any language" and called Henry Vandyke Carter's illustrations "perfect." Indeed, they are perfect, both exquisitely wrought and functional. His great innovation was to place the anatomical names right on the parts themselves, like street names on a road map—in spots, terms even curve right along with the anatomy—something students found enormously helpful. By comparison with Carter's originals, the added drawings look blunt and diagrammatic. The most striking difference, however, is the first edition's lack of color, which I am surprised to discover I favor. Carter's drawing of the man in profile— the image that first captured my attention—is more beautiful in the original, where it appears not on the cover but a third of the way into the text. Seeing it as Carter intended is like seeing a masterpiece restored. And the colored version—*how have I not noticed this before?*— now looks garish.

Not surprisingly, I find subtle text differences between the 1858 and 1901 editions; in the latter, words are substituted, sentences shortened, punctuation changed. As a result, the book, already clinical in tone, was made even chillier. In the original, Henry Gray often provided brief introductory remarks for each section, which set a welcoming tone. Forty-three years later, his remarks were gone.

One thing is exactly the same in both editions: the book comes to an abrupt conclusion. It is almost as though Professor Henry Gray, in the midst of lecturing, sees that he has gone past his allotted time. His words quickly grind to a halt—". . . and receives a prolongation from it." And that's that. Class dismissed. In the first edition, however, two last words appear, in tiny print:

THE END.

As I sit in the library, those two little words sound wonderfully ironic. Could Henry Gray ever have imagined what "The End" would begin, the long life his work would enjoy? Having never gone out of print, it has to date seen thirty-five editions in the United States alone. It has been translated into more than a dozen languages, been pored over by generation after generation of medical students, and sold millions of copies.

I try to imagine what was going through his head when, early in 1858, Henry put the finishing touches to his tome. I picture him sitting at a meticulously organized desk in the Gray family home on Wilton Street, where he lived alone with his widowed mother. The hundreds of handwritten manuscript pages are stacked in a neat tower, ready to be boxed up for his publisher, when the thirty-one-year-old gets bitten by whimsy. He pulls out a fresh sheet of paper and, with his most careful calligraphy, writes those two last words. He slips this final page into the bottom of the stack. He does not expect it to survive the editing process; this is his attempt at a little joke. *"The End"? Yeah, right.* A book ends, a story ends, a life ends. But the desire to study anatomy never will.

Two

IN BOOKS ON HUMAN ANATOMY, THE SKELETON IS GENERALLY either the end point or the starting point. The two Henrys, Henry Gray and Henry Vandyke Carter, chose the latter for their tome. "In the construction of the human body," the text begins, "it would appear essential, in the first place, to provide some dense and solid texture capable of giving support and attachment to the softer parts of the frame; such a structure we find provided in the various bones, which form what is called the Skeleton." Gray's tone here, eminently reasonable and deceptively conversational, immediately draws the reader in.

The first drawing one sees is of a cervical vertebra, one of the seven bones that, stacked atop one another, form the skeleton of the neck. Typical of Carter's style, the bone is rendered with a fine, delicate line and with perfect shading to show depth and dimension. Though elegantly drawn, I somehow doubt it is the kind of thing his mother had in mind when, years earlier, she dreamed that her firstborn son would become a famous artist. Eliza Carter reportedly had so hoped that Henry would follow in the footsteps of the great

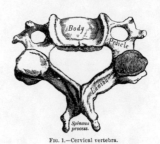

FIG. 1.—Cervical vertebra.

seventeenth-century Flemish painter Sir Anthony Van Dyck, known for his lush portraits of English royalty, that she chose Van Dyck as his middle name. On the day of young Henry's baptism, however, the desired spelling was misentered into the parish registry as "Vandyke," an error that endured. Regardless, he would come to use his full given name only rarely, preferring to go by the initials "H.V."

Carter may have inherited his mother's hopes, but his artistic talent came from his father, Henry Barlow Carter, a popular Yorkshire watercolorist known for his landscapes. In addition to H.V., born on May 22, 1831, the Carters had a daughter, Eliza Sophia, called Lily, and a second son, Joseph Newington. (Lily and Joe, born in 1832 and 1834, respectively, shared the same birthday, December 26.) The family lived in Scarborough, a seaside village in northeastern England, where Mr. Carter was an art instructor and artist in residence at the local library. What provoked the young H.V.'s left-hand turn toward medicine is not certain, but two strong influences have been identified. His uncle, John Dawson Sollitt, was the headmaster at his grammar school and possessed a keen interest in science, as did one of H.V.'s older cousins, yet another Henry—Henry Clark Barlow—who was a physician as well as, curiously, a Dante scholar. Following his completion of grammar school (the British equivalent to high school), Carter, fifteen, became an apprentice to a pair of Scarborough physicians and during these nine months learned the rudiments of country medicine. A young man, however, did not become a licensed practitioner in a sleepy place such as Scarborough. At sixteen and a half, H. V. Carter "came to town," as he would later put it, moving by himself to London, the largest city in the world at that time and home to a number of medical schools.

Unlike a student entering medical school today who steps onto the educational equivalent of a moving sidewalk—a set course of study, logical and well organized, leading straight toward a medical degree—Carter had to follow an often circuitous path in pursuit of training. Still, at least a path had been paved. Just a generation before, there were no established guidelines for a young man seeking a medical profession, even in the influential city of London. Writing of that period, British medical historian Charles Newman notes, "The process was entirely unorganized—it was left to the student to decide on his own curriculum and to find out how it could be followed." While improvements had been made by the time of Carter's arrival, the system remained disorganized, albeit in different ways. Now, seventeen independent licensing bodies existed—the Royal College of Surgeons, the Society of Apothecaries, the Royal College of Physicians, and so on—each with its own accreditation criteria.

Carter's father had arranged for his son to be placed with the Royal College of Surgeons, under whose purview H.V. became an "articled student of medicine"—that is, apprenticed—to a London doctor, Dr. John James Sawyer. This was a legally binding agreement, for which Carter's father had to pay a fee of ten guineas to the RCS. In addition to on-the-job experience with Sawyer, H.V. would live with the doctor and his family over the next three and a half years.

Freshly articled and newly settled, H. V. Carter was then able to take his next big step in becoming a doctor: on May 27, 1848, he registered as a student at St. George's Hospital Medical School and immediately plunged into full-time coursework. Coincidentally, as Carter was starting his education at St. George's, Henry Gray, four years older, was in the last year of his. Though their momentous collaboration was still a decade off, it's safe to say that the two men first met, at least in an academic context, in the last weeks of 1848. As was true of Mondino's time, dissections were performed only during the coldest months, so Carter's study of the human body did not begin until the start of the winter session. And, as fate would have it, Henry Gray had just been newly appointed as demonstrator of anatomy.

So did the two become fast friends? When did Gray learn of Carter's artistic talent? Was the *Anatomy* their first work together?

Well, of course, Gray doesn't say. While the historical record for the famed anatomist is silent, such is not the case for his lesser known colleague. In fact, in the time leading up to my first day of anatomy class, I had discovered that a trove of H. V. Carter's diaries, letters, and other personal documents was stored at the Wellcome Library in London. The papers, which date from his grammar school days to the end of his long life, had scarcely been studied. With tact and a credit card, I was able to persuade the library's archivist to have the first two diaries microfilmed for me, sight unseen, so that perhaps I, too, could witness life in London in the middle of the Victorian era and, through H. V. Carter's eyes, hopefully get a glimpse of the inscrutable Henry Gray.

Only after I had placed my order did a sinking feeling hit: What if I could not read Carter's handwriting? That doctors have notoriously bad penmanship can hardly be unique to our day and age. What if the diaries were impenetrable, and that is why Carter's story remains largely untold? Just as worrisome: what if he had used his diary not as a repository for his feelings and experiences but as a mere date book, filled with nothing but class notes and study schedules?

Six long weeks later, my answer arrived by mail on a fat spool in a sturdy square box. I headed immediately to that last refuge of the antiquated technology of microfilm, the public library. Providing tech support and, if needed, moral support, along with me came my longtime partner, Steve.

Steve fed the thick, wide film leader into one of the brutish old projectors, and I, with fingers crossed, pressed the Forward button. First came the loud flapping sound as the microfilm struggled to catch onto the receiving spool, then the quieter hum as it sped through the projector. A long stretch of velvety blackness filled the screen and then a blinking brightness. I backed up to page 1.

The diary of H. V. Carter got off to a very promising start. In the opening lines, written in a large, childlike script, I got immediate answers to my first questions: why and when did he start keeping a diary?

A gift from his "Grandmamma," it reads, the diary was "to be commenced May 22nd, 1845," the boy's fourteenth birthday, "when leaving Scarborough for school in Hull." (Hull was a city down the coast where he would be a boarder at the grammar school headed by his uncle.) What immediately follows this text, though, is not the musing of a fourteen-year-old but instead a terse disclaimer written by Carter seven years later:

"I began my Journal at the above date or soon after and continued it for at least six months being then at School in Hull," he explains, "but from an unpleasant occurrence happening at this time, the journal was altogether discontinued—the existing pages being destroyed—and was not resumed till the end of '48 when I came to town. Since then, with but one exception," he adds, "I have kept a continuous daily record. . . ." His handwriting in this portion is barely legible—small and cramped as if, in the intervening years, the young man had folded in on himself.

Exactly what unpleasantness occurred Carter does not reveal, but, hazarding a guess, perhaps some school bully found his diary and threatened to divulge his secrets. This gave me pause. Was I violating H. V. Carter's privacy, under the guise of research? At the same time, I felt that he could not have a more sympathetic reader. I, too, had started a journal at age fourteen only to rip it up a few months later. By destroying the pages, I could almost believe that the sinful thoughts I had recorded would cease to exist. I could then start fresh, my soul a blank white page. But like the young H.V., I continued to write. Over the years, I filled notebook after notebook and kept them as well hidden as I had learned to hide my inner self. In fact, I kept journals until my need to keep secrets finally ended when I came out in my early twenties. And yet, two decades later, I still have the journals, every last one.

Moving to the second page of Carter's diary, I found him, precisely as he had noted, in London in December 1848, a seventeen-year-old halfway through his first year of medical school. From here, page after page of daily entries form weeks, then months, then years. Whirring through the microfilm, I stopped every now and then, like a crow drawn to a shiny object, and picked up pieces of his story. I found bright bits, but dark ones, too, admissions of success but also of failure and sin. His handwriting was sometimes loose and legible, but most often it resembled long strings of tiny knots. Here and there, familiar names and places popped out. Most exciting to see was how, beginning in 1850, the black-on-white pages became sprinkled with *Grays*, the name always written in beautiful cursive. I had found what I had been hoping for—Henry Gray lived on these pages—but there was more. The sprawling paper trail left behind by H. V. Carter would lead me not just through the winding corridors of St. George's and into the dissection lab on nearby Kinnerton Street but, most intimately, most tellingly, deep into the troubled heart of a gifted man of science.

"YOU'RE BACK?" MASSOUD says when I join him at the dissection table on day two of class. His dark, bushy eyebrows have raised to the point of looking painful. "I cannot believe you'd come here voluntarily."

I laugh and admit he has a point—most people would not choose to spend an afternoon disassembling a body, a gruesome business made more so by the harsh embalming chemicals. But to me, this is a small price to pay for seeing the extraordinary, the inner architecture of the human form.

Massoud and his classmates, obviously, do not have a choice about whether to be here. Each must pass this anatomy course in order to graduate, as must those enrolled in UCSF's dental, physical therapy, and medical schools. As to why it is mandatory for pharmacy students, that is easy to understand. To grasp the basics of how medications work within the body—from, for example, the placing of a pill on the tongue to its passage down the throat and course through the digestive, then circulatory systems—one must first grasp the fundamentals of how the human body is constructed. Hence, ten weeks of Gross Anatomy, *gross* coming from the German for "large" and referring to structures of the body that are visible with the naked eye.

Dr. Rohde approaches our table and, before even saying a word, instantly captures our attention: she is holding a human heart. "The most *amazing* thing we do as human beings occurs the moment we're born," she goes on to say. "We have to learn how to breathe on our own." And for the rest of our lives, the heart bears a scar of this life lesson. One of the goals for this afternoon's lab is to find this mark in our cadaver.

Massoud and I take turns reading aloud from the lab guide as Gergen, Laura, and the three other students undrape the body and align the dissecting instruments. To my surprise, Amy, the least assertive of the bunch, agrees to perform the dissection. Amy is just over five feet tall, stocky, with bobbed brown hair and funky rectangular eyeglasses. For added height, she steps onto one of the wood risers positioned around the table, then picks up a scalpel and, following our instructions, makes a large, neat "cruciate cut"—cross-shaped—atop the pericardium, the opaque protective sac that encloses the heart and helps hold it in place. The pericardium is composed of multiple layers, with the final, thinnest layer adhering ever so lightly to the organ itself.

Amy slides her finger into the center of the cut and folds back each flap, exposing the heart. She reaches for a larger blade. Amy looks so comfortable using a scalpel that I cannot resist asking if she's ever thought about being a surgeon rather than a pharmacist.

"Not until now," she answers with a smile.

Next, Amy slices through the six blood vessels entering the heart and the two exiting it. Then she puts down the knife, grasps the heart with both hands, and tugs, uprooting the organ from its bedding in the chest. She places it on a towel-lined tray to her left.

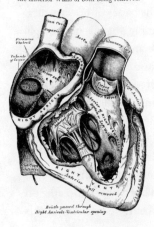
328.—The Right Auricle and Ventricle laid open the Anterior Walls of both being removed.

The human heart has four chambers, but it is not obvious from the outside where each is located. Six noses press in for a closer look. Subtle grooves on the exterior of the organ serve as landmarks, and we are able to orient ourselves. The right and left atria, as their names suggest, are the two cavities up on top. The right and left ventricles form the lower portions of the heart.

Amy proceeds with the final incisions. Using a fine blade, she makes a small doorway into the right atrium and, turning the heart over, a larger opening in the left ventricle. Our heart now has a front and back door, but there also seems to be a flooding problem. Dr. Rohde—Dana, as she insists we call her—has been observing how we are doing, and she suggests that someone take the heart to the sink and rinse it out.

I volunteer.

With an air of quiet ceremony, Amy places the heart into my gloved hands, and I instinctively draw it to my chest. My own heart instantly speeds up. The lab has never seemed more crowded, the distance to the big stainless steel sink never more vast. I feel as if I were carrying the most fragile thing in the world, which is silly, for our heart is already broken in a sense; our cadaver had died of heart failure.

Once I begin rinsing the heart, cradling it in one hand while rubbing it with the other, I relax. It is tough and rubbery. The aorta, the major artery emerging from the heart, is a severed garden hose. As I feel the smaller vessels, white and gristly like the roots of a turnip, I understand how the word *heartstrings* came to be, based as it was on the belief that stringlike tendons keep the heart in place and can be tugged or plucked like harp strings, eliciting different emotions.

What washes down the drain is a grainy brown paste, coagulated blood from inside the heart. I pat the heart dry and return to our table.

With Dana as our guide, we examine the four chambers in the same order blood passes through them, beginning with the right atrium. This is where blood is received from two veins—carriers of deoxygenated blood—the superior vena cava and the inferior vena cava. (*Superior* means topmost and *inferior,* bottom, terms that we would come across again and again.) The right atrium pumps blood into the ventricle beneath it, which pumps it into the lungs. Blood returns to the heart via the pulmonary veins (the only veins in the body that carry oxygenated blood) and enters the left atrium, which pumps it to its partner below. The wall of this last chamber, the left ventricle, is the thickest and strongest of them all; it has to be. With each pump, the left ventricle propels blood up through the aorta and out through the body's miles and miles of arteries.

But before supplying blood to the whole body, Dana notes, the heart does something very wise. "Does anyone know what it is?"

She is answered by six shaking heads.

Dana points to two slim vessels emerging from the base of the aorta and snaking down the heart's surface. These are the right and left coronary arteries. In what makes perfect sense now that it's pointed out, the first destination for fresh, oxygen-rich blood is here. "Remember this phrase," Dana tell us: " 'The heart feeds itself first.' "

Remember it? I'd like to contemplate it. But there is no time. Dana and the team are already focused on the right atrium. Through Amy's doorway, they find the scar Dana had mentioned earlier, a thumbprint-sized indentation.

"In utero, this used to be a hole," Dana says. During fetal life, blood passes not into the lungs but directly from the right to the left atrium through this shortcut. Though the baby is not breathing in a technical sense, it is getting plenty of oxygen, drawing it from the mother's bloodstream through the placenta.

"But the shunt becomes obsolete at birth," Dana continues, "when a newborn, gasping for air, uses its lungs for the first time." This single act radically changes the pressure within the circulatory system, channeling blood *into* the lungs rather than away from them. No wonder a newborn howls. Within hours, the hole begins closing up, leaving behind this fossil of fetal life, the fossa ovalis. In some babies, Dana adds as we each take turns examining it, the shunt does not heal properly, leaving an actual hole in the heart, which has to be surgically repaired.

You cannot hold a human heart without questioning how it ever became known as the center of emotion or, as the grand sixteenth-century French surgeon Ambroise Paré once described it, "the chief mansion of the Soul, the organ of the vital faculty, the fountain of the vital spirits." To me, the heart does not look or feel like anything but what it is, a tough, muscular pump. But wait, not so fast.

"Let me show you one last thing," Dana says before moving on to the next group and the next body.

Massoud, Amy, and the rest of us crowd around her as she lifts u
the cadaver's heart and pulls the doorway into the right atrium as fa
back as it will go.

"Now, unfortunately, you can't actually *see* it," Dana says, "but,
right inside *here*, where the superior vena cava enters the right
atrium"—she points to a spot at the top of the fold—"right at that
ridge is a little area where a cluster of cells is embedded. It's called
the sinoatrial node, or S-A node, but it's known as the *pace*maker."
She lets that sink in. "This is where your heart's speed is set."

While she explains how the S-A node works—electrical signals
generated by these cells spread to other cells across the heart, caus-
ing it to contract, to beat—I find myself dazzled by this perfect meeting of anatomy and metaphor. In the human body, the node is
positioned right under the sternum, dead center in the chest. So, in a
sense, this truly is where feelings such as terror, love, and elation are
first felt—where your heart starts to race, pound, flutter.

Looking up, I notice that Amy is doing exactly what I am doing:
we both stand with a hand at the center of our chests, instinctively
feeling the moment. Here, right here, is where wonder begins.

Three

"I HAVE TO SAY, KIDNEYS ARE ONE OF THE *SADDEST*-LOOKING creatures!" laments Dana during a lab presentation midway through the ten-week course. I have to agree. The sickly gray organ, which she had just removed from a demo cadaver's lower back, looks pockmarked, blob-shaped not kidney-shaped, and, indeed, sad. Though larger, it reminds me of a testicle, or at least the testicles we had studied a couple of weeks earlier. Do I detect a family resemblance?

Indeed, in males the two *are* connected, not directly but venously, Dana goes on to explain. Bridging the twelve or so inches from the left testicle to the left kidney is the testicular vein, which feeds into the renal vein as blood returns to the heart. This particular anatomical arrangement occurs only on the left side.

"In fact"—Dana cracks a playful smile—"in fact, this may explain why the testes hang unevenly."

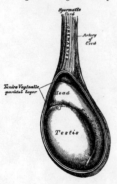

343.—The Testis in Situ. The Tunica Vaginalis having been laid open.

Spermatic Cord

Artery of Cord

Tunica Vaginalis parietal layer

Head

Testis

The men in the group share a nervous chuckle.

"You know what I'm talking about," Dana says matter-of-factly. "Usually, the left testicle hangs lower than the right. Right?"

One could almost see the wheels turning as the assembled males each perform a mental inspection of their underwear.

"Right," I volunteer on behalf of my shyer classmates.

"Well," Dana continues, "this is likely

because the left renal vein runs between two high-pressure arteries, so it may get slightly occluded—or squished. Less blood can travel through it, so the blood pools down in the left testicle, making it a little heavier than the right one."

Now here's some juicy small talk for a lagging dinner party, I can't help but think.

"Okay, let's move on now," Dana chirps, drawing our focus back to the object in her gloved hand. The "pitiful appearance" of kidneys is deceiving, she notes; these are strong, resilient organs, capable of impressive multitasking. They not only filter waste and toxins from the blood but regulate urine excretion while simultaneously maintaining the body's electrolyte and fluid balance. If one kidney is removed or fails, the other will pick up the slack and do double duty.

The kidneys also provide a perfect illustration of an age-old anatomical truth: the body is designed to protect itself, not to be easy to dissect. As Henry Gray noted in his precise fashion 150 years ago, the kidneys are situated between the back of the abdominal cavity and another eight separate structures, including two powerful back muscles, and are "surrounded by a considerable quantity of fat," even in the lean. All of which makes finding a kidney in a cadaver tricky. The best method, Gray advised, is not to go through the abdomen but to flip the body over, count down to the last rib, drop down another three-quarters of an inch (about two centimeters), then make your incision there.

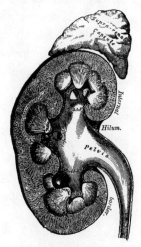

Before rotating to the next group, Dana shows us the proper way to open up a kidney, carefully splitting it in two lengthwise. Like a pomegranate, whose leathery rind belies its jewel box interior, the kidney is spectacular on the inside. Each half is lined with the small chambers and pyramid-shaped tissue of the organ's filtering system. Once everyone has taken a close look,

FIG. 543.—Vertical section of kidney.

we break up into our smaller groups and return to our cadavers. The goal: to replicate what Dana had so nimbly demonstrated.

As an observer, I have the option to move about the lab, from cadaver to cadaver, from group to group. (This is how I'd finally met the Woman in the Gas Mask from the first lab. Beneath the hazmat headwear was a lovely person named Iris, who is pregnant, it turns out, and, on the advice of her obstetrician, takes the extra precaution for her baby.) Although each lab lasts three hours, students are free to leave as soon as they finish the day's assignments, and most of them do. But I like to stick around until the last body is re-draped. The whole experience has quickly come to seem normal to me; friends beg to differ, however, when I mention I am attending a course in anatomy.

"You mean, with bodies?" is always their first response. "Actual dead bodies?!"

What's missing from their mental picture, I have come to understand, is the larger context. Just as a person who has never before stepped inside a church could gather from the altar and hushed candlelit atmosphere that it is a place of worship, so, too, could one enter the anatomy lab for the first time and readily grasp its purpose. Chalkboards line the entire back wall. Bookstands, poised at every table, hold identical manuals. Display cases and neatly labeled drawers contain anatomical models and specimens. Most important, though, is what happens about ten minutes into each lab: the instructors enter, at once transforming the space into a learning center of crackling vitality.

In putting together a team for the course, Dana's first move was to coax back from early retirement the man she considers one of the leading anatomists in the United States, Dr. Sutherland. Tall and lanky, with silky white hair, Sexton dresses for comfort in sneakers and khakis and always wears whimsical neckties—one has dancing skeletons on a blood-red field. The antithesis of a dour anatomist, Sexton is sunny and self-deprecating, and in the lecture hall, a bit of a klutz, which is actually quite endearing. His clip-on microphone often falls off; he has trouble finessing the overhead light dimmer; his slides sometimes come up sideways (we all tilt our heads obligingly). The man obviously knows anatomy backward and forward—or, forgive me, posterior and anterior, as well as medial and lateral, superior and inferior, and in every other anatomical position—but he also makes it entertaining. In summing up the core behavioral impulses regulated by the sympathetic nervous system, for instance, Sexton once told the class: "Just remember the four Fs: Fight. Flight. Fear. And—who knows the last F?"

Puzzled silence.

"That's right," Sexton said with a knowing nod. "Sex!"

Sexton brings the same exuberance to the lab, where, like his fellow instructors, he roams from group to group, answering questions and giving impromptu lectures. Each teacher has a different style. Dr. Nripendra Dhillon—Dhillon, for short—is the third of the trio of senior instructors and a master of visuals. I mean this both literally— he will often sketch on any nearby chalkboard, whether in the lecture hall or lab—and metaphorically. Lecturing on the intrauterine development of male reproductive organs, for instance, Dhillon made the descent of the testicles through the fetal body sound as dramatic as Odysseus's epic journey home from Troy. With his deep, melodic voice, Dhillon recounted how the testes actually develop in a pocket of fat on the fetus's back, behind the kidneys. But at around the ninth week of fetal life, these delicate little, well, *balls* ship off. Traveling separately but to a similar map, they slowly traverse the lower abdomen, pushing through layer after layer of abdominal tissue, acquiring new coats as they tunnel to their final destination: the scrotum. Though any man who has been kicked in the groin might not think so, these added layers actually provide protection. To make sure this journey was ingrained in our memory, as Dhillon spoke, he pulled successive, colored latex gloves over his right hand to represent each new layer—purple, green, pink, and finally, blue—each time balling his fingers into a thick, rubbery fist.

Two teaching assistants round out the team. Because Christy and Aaron were so recently students themselves, they are especially helpful in sharing mnemonics and other time-saving study tips. Of all the instructors for this course, though, Dana herself has made the strongest impression on me. In what I take as the highest form of flattery, she never treats me like an observer but as one of the 121 students in the class, even grilling me good-naturedly in the oral pop quizzes she sometimes springs during lab. Given her obvious enthusiasm for the subject of anatomy, I was surprised to learn, though, that Dana had never set out to become an anatomist.

"I'm definitely an 'accidental anatomist,'" she told me one afternoon as we chatted on the way up to lab. After earning a B.S. in nutrition, a master's in biology, and a Ph.D. in physiology, Dana had planned to go straight into medical research. But there was a surprise on the menu: an offer of a teaching job landed on her desk—UCSF was seeking a physiology instructor—and it was unexpectedly tempting. She accepted and was amazed to find how much she enjoyed teaching. Then she met Sexton and realized she would also like to teach the subject her new friend was so passionate about. He thought this was a fabulous idea; in fact, the anatomy department had an opening coming up. But first, Dana would have to turn herself into a great dissector. Sexton became her mentor. He spent hours and hours of extra time helping her learn how to perform the most difficult dissections. The greatest lesson he taught her, though, was one of aesthetics: how to make dissections beautiful.

"For a year, I was here *all* the time dissecting," Dana said once we had reached the thirteenth floor, "even every Saturday night. That's the way you learn anatomy. You sit down with a dissection manual and a cadaver, and you just slowly go through everything."

I WOULD LOVE to have been an observer as Henry Gray learned the art of dissection. Colleagues who remembered him as a student invariably recalled a "most painstaking" and "methodical" worker but left no more telling details or anecdotes in the historical record. Fortunately, however, I am able to reconstruct the setting where the young Mr. Gray spent hundreds and probably thousands of hours quietly following his passion.

A fifteen-minute walk from his family home on Wilton Street would have brought an eighteen-year-old Gray to the north end of Kinnerton Street, to the building where St. George's anatomy courses were taught. In what he and his classmates probably found a dubious comparison, the premises were often likened to the inner architecture of the human ear. The building was set well back from the street, and, just as the ear canal leads to the eardrum, one passed through a long, narrow alley before reaching the main door. Completing the analogy, the school's circular anatomical theater represented the spiraling cochlea at the innermost part of the ear. Lecture halls, an anatomical museum, and an impressive dissection lab rounded out the floor plan.

That Kinnerton Street was a good four blocks from St. George's Hospital was seen by the medical school administrators as a significant drawback but also a marked improvement over the previous accommodations. When the hospital established its school in 1829—sixteen years before Gray's enrollment—the board of governors declared that anatomy would not be taught on the hospital's grounds themselves but close by. Right across the street, in fact. An excellent independent anatomy school had just opened and could easily accommodate the St. George's students. What could not have been foreseen, however, was how one of the directors of the anatomy school, James Arthur Wilson, would constantly come to loggerheads with hospital administrators. A bilious-sounding character who went by the nickname Maxilla (the anatomical term for the upper jawbone, inspired by his initials, J.A.W.), Dr. Wilson was described with admirable delicacy by one historian of the period as a man "somewhat over-conscious of his own excellencies." After one too many quarrels with Maxilla, St. George's chief surgeon, Benjamin Brodie, ended the association between the two schools and, in 1834, financed the purchase of the Kinnerton Street facility.

By the time Henry Gray began attending classes there in 1845, Dr. Brodie had retired as surgeon and anatomy instructor but, at age sixty-two, continued to practice medicine and was regarded as one of England's leading medical authorities. As writer James Blomfield observes in his history of St. George's Hospital, Brodie had attained a degree of public acclaim rarely seen nowadays: "It is difficult for those who live in the present day, with specialists for every kind of complaint, to imagine the position of a man like Brodie. He was consulted by patients of all ages and upon almost every conceivable form of accident or disease." One famous case involved a gentleman who, in a conjuring trick gone awry, had accidentally inhaled a half-sovereign coin, which then lodged in the man's upper right lung. However, it seems that the true performance did not start until Dr. Brodie's arrival. Immediately, he turned the patient, Mr. Brunel, upside down, a feat made relatively easy by the man's owning a "revolving frame," which I will assume was like a knife-thrower's prop: a circular board to which, under normal circumstances, a lovely assistant would be strapped at the wrists and ankles then spun while knives are hurled. Upside-down-ness did not, however, help Mr. Brunel cough up the coin. Rather, the object plugged his larynx and he began to choke. With a confident slice of a blade, Dr. Brodie opened the man's windpipe but, even with forceps, could not dislodge the half-sovereign. Another spin on the frame, however, did the trick. Gravity, a smack to the back, and a fortuitous gag reflex caused the coin to drop quietly into the man's mouth. In tribute to the doctor's calm under fire, the half-sovereign and the pair of forceps became one of the exhibits in the St. George's Pathology Museum. As for Mr. Brunel, I can only hope he had the good sense to move on to card tricks.

Though he had left behind his role as instructor, Brodie maintained a keen interest in the medical school he had helped found, and, through one channel or another, word of the talented Henry Gray came to his attention. The most likely messenger was Brodie's nephew-in-law, Thomas Tatum, one of St. George's top surgeons and an anatomy instructor for almost twenty-five years. That Brodie and Gray met is a certainty, but when? Interestingly, an answer is suggested in a dinner invitation that survives to this day—Sir Benjamin and Lady Brodie inviting Henry Gray to their home on Monday, the twenty-eighth of April—though the year is uncertain. A little detective work tells me that this day/date combination occurred only three times during Gray's adult life—in 1845, 1851, and 1856. Of the three, the first date offers the most intriguing possibilities. Monday, April 28, 1845, is eight days before Gray registered at St. George's medical school. I find supremely satisfying the idea that this is when he first met Benjamin Brodie, the legendary man to whom, thirteen years later, he would dedicate his great work, his *Anatomy*. I picture an intimate gathering, with Dr. Brodie personally introducing Henry to a few distinguished colleagues, the young man's eyes as round as the Wedgwood plates as he shook hand after hand. But why would Sir Benjamin and Lady Brodie have invited this young nobody to their Savile Row residence? Well, it turns out, Gray had won a prestigious "junior prize" in anatomy as a sixteen-year-old, and the lad's burgeoning talent had clearly impressed Dr. Tatum. Indeed, it was Tatum who, the following week, would cosign Henry Gray's registration as a medical student.

But there's a final reason I hope this early date was in fact their first meeting, for the warm invitation would serve as a prologue of sorts to Gray's career just as another note from Brodie would serve, sixteen years later, as a fitting epilogue. Upon receiving news of Henry's sudden passing, Dr. Brodie, at age seventy-eight and in failing health, wrote to a colleague: "I am most grieved about poor Gray. His death, just as he was on the point of obtaining the reward of his talents, . . . is a great loss to the Hospital and the School.

"Who is there to take his place?"

HENRY VANDYKE CARTER prepared for the first day of his first year of dissection in the same way a student today would: he shopped. After taking a quick look around the new laboratory at Kinnerton Street, the eighteen-year-old went and placed an order for a dissecting "gown," a kind of loose cassock (a precursor to the green cotton scrubs of today), and then headed to Savigny & Co. and bought a "case of scalpels," he reports in his diary on Saturday, September 29, 1849. Carter could not afford a copy of the standard anatomy guide,

Quain's *Elements of Anatomy*—"Funds low," he notes in his usual clipped style—so he would just have to make do without.

The winter session would begin with speeches and an awards ceremony on Monday, and lectures and lab work on Tuesday. Carter, who had spent the past year and a half sitting through classes in anatomy, botany, physiology, chemistry, materia medica, and medical jurisprudence, would finally get his hands bloodied. "All prepared," he writes before bedtime Monday night.

But it takes two to take the next step. "Not dissect for subject not ready," he writes the following day ("subject" meaning cadaver), which may have been for the best since Carter's gown was not ready yet either. Finally, on Wednesday, October 3, he makes his debut as an anatomist. Under the watchful eye of Dr. Athol Johnson, he begins with a part of the body both relatively simple to dissect and, if you picture Michelangelo's *David* as the ideal, lovely to behold: the inguinal canal, the area where the lower abdominal muscles slope down toward the groin. At the close of the day, Carter confides: "[I] like dissecting. More difficult than [I'd have] thought without guide."

This last little admission is the kind of ironic detail that brings a smile to my lips, knowing as I do the role H. V. Carter will go on to play in creating the most famous anatomy guide of the past two centuries. Another such moment comes three pages later, with the first mention of Henry Gray. So synonymous has the name *Gray* become with *anatomy*—as familiar a pairing as *Webster* and *dictionary*—that it is jarring to see it spelled incorrectly, as Carter does on October 31, 1849. The error is unusual for him, an impeccable speller otherwise, and suggests that the two men did not know each other well yet. As for the mention—"See Grey, promise" is all he writes—it makes no sense to me. But that's fine; it is part of the odd dynamic that develops between diarist and reader as the lopsided omniscience borne by both gets traded back and forth. Which is to say that at any given moment, on any given day, Carter experiences far more than he ever puts into words, just as I, on any given page, know far more than he about the course his life will take.

For Carter, keeping a diary had been intended originally as a character-building exercise, a good "habit" for a young man to keep (good habits being prophylaxis against bad ones). For me, deciphering his diary has been like performing a dissection in reverse—a slow piecing together. The process has required spending numerous hours in the microfilm reading room at the library, where Steve and I have gotten to know the tics and quirks of each microfilm projector and become familiar with the microfilm-reading regulars. Those of us who gather there are an odd little community of time travelers. Most everyone reads newspapers, an old-fashioned pastime made more so by the age of the newspapers themselves—antique issues of the London *Times,* say, or a monthlong run of the defunct *Chicago Herald.* Steve and I are not so different. The daily news we're reading just happens to be in the form of a diary.

In what now reads like an epigraph, Carter started his diary with an aphorism that sounds as if it were taken from a Victorian-era self-improvement book: "Let the same thing, or the same duty, return at the same time everyday, it will soon become pleasant." He would frame each entry, beginning with the time he woke, closing with his bedtime, and capturing the hours in between in a few deft lines. Just as a painted portrait acquires depth and texture with the accretion of paint, an image of H. V. Carter emerges only after many weeks of entries. A serious, disciplined young man, he reads the Bible and prays daily, and goes to church—often twice—on Sundays, but he is also only seventeen years old and had moved from a town of ten thousand to one of more than a million, so naturally a boyish excitement bursts through every so often. He is left almost speechless one day by a sighting of Queen Victoria, while a few weeks later, he is fascinated by troops practicing formations in Hyde Park. At the same time, his eyes are also being opened to unpleasant realities. On May 23, 1849, the day after his eighteenth birthday, Carter begins serving a "clinical clerkship" at the hospital, a position in which he would shadow staff surgeons and take their case notes. Just two days later, he witnesses a horrific procedure, the amputation of a boy's leg. "Chloroform not used," he writes that evening, which comes as a chilling reminder that anesthesia during surgery was not yet standard practice.

At times, Carter's prose is so immediate and concise, it is as though he were dictating a telegram. "Cholera case," he writes on July 6. "First I've seen. Came in yesterday 6:30 P.M. Died 6:30 P.M. today. Terrible disease." The next day, he attends the postmortem examination of this patient and is stupefied; the man doesn't look dead enough to be dead. (Cholera, a bacterial disease spread mainly through contaminated drinking water, causes devastating diarrhea and dehydration.) By August 1, the cholera outbreak has so overwhelmed London's hospitals that Carter must watch as St. George's shuts its doors to new patients.

Contrasting with entries on the latest death tolls are warm passages on his life in the home of John Sawyer. Sawyer, forty-five, the same age as Carter's father, ran a private practice and apothecary out of his Park Street residence, and he and his wife had five daughters, ranging from nineteen years old down to six. Although Carter paid an annual fee for room and board, it is clear that he was embraced as a part of the family and that this home away from home provided much solace. On Sundays, he often joined the Sawyer family on walks through the nearby parks and spent the afternoons playing the role of indulgent big brother to the younger Sawyer children. And on Sunday evenings, he oftentimes accompanied the second eldest daughter, Mary, to chapel. Carter always comes across in his diary as a polite and proper young gentleman, never mentioning anything untoward, yet his hormones were definitely speaking to him. "Avoid temptations," he includes in a list of gentle reprimands, and, "Be careful to improve your thoughts when alone." If idle hands are the devil's tools, as the saying goes, then the devil was two mitts shy once winter session 1849 got under way.

Seemingly overnight, Carter's diary turns into a chronicle of anatomizing. Not only does he dissect in class most days of the week but sometimes at home as well, using souvenirs, for lack of a better word, he had gotten at postmortems. "Got two eyes," he reports one night, obviously pleased, as if *one* eye would have been a big disappointment. "Got kidney and heart," another day. And, once, "Had offer of brain, but declined," a rare demurral. He also obtains parts from the hospital's morgue, the aptly named Dead House. Somehow, though, his hunger to dissect never sounds ghoulish. To read Carter's entries is to watch a young man chasing after knowledge at full tilt. He misses lectures because he loses track of time in the lab. He works through lunchtime, missing out on eating. From lecture to lab, lecture to lab, he sometimes returns to Kinnerton Street three times in a day. Ever fastidious, Carter will often record how long he spends dissecting, as if he were a runner training for a race, pushing himself to beat his own record. Though he takes off Christmas Day, he is in the lab New Year's morning.

Carter's growing mastery of dissection does not go unnoticed. Ten weeks into the session, instructor Prescott Hewett asks him to make a "preparation" for the anatomical museum. In other words, he would dissect some body part, which would then be bottled and preserved in "spirits" (alcohol) for students to study for years to come. In general, a preparation would have been done by a faculty member, but Carter was obviously gifted. And excited! For three days, he nervously awaits word of his assignment. He is given a hand, it turns out. "With <u>name</u> to be added," he writes, meaning his name will be affixed to the bottle for posterity.

With the preparation turning out well, Dr. Hewett presents his protégé with a copy of Quain's *Anatomy*, a much-appreciated gift. Carter inscribes his name in the book and, on his way home, purchases a protective cover for it. His pleasure overflows to the next day, when he pages through the illustrations and paints all the arteries red. This "Q" is "fine work!" he writes.

By this time, January 1850, Henry Gray had been promoted from demonstrator of anatomy to the hospital's postmortem examiner. The twenty-three-year-old had also just enjoyed the honor of having a paper read before the Royal Society. That he was so rapidly making a name for himself made a deep impact on Carter, as shortly becomes clear in the younger man's diary: "Must work!" he admonishes himself. "Gray getting on!" It is as though Gray were the pace-setter, and Carter, following a similar career path, does not want to lag behind for a moment.

That same day, Dr. Hewett had offered him a new project: to pre-serve some anatomical specimens not by dissecting but by drawing them instead. The subjects would be hospital patients with unusual maladies. Keen to help out, Carter agrees, and, two weeks later, he finds himself in the women's-only "nurse's room," performing the delicate task of painting a woman's diseased breast. Carter "man-age[s] tolerably" and finishes in an hour, after which the breast is sur-gically removed. His next subject is a thigh of unspecified pathology.

Nowhere in his diary do I get the impression that Carter ever imagined he would be bridging medicine and art. True, his father was a working artist and he had grown up drawing and painting, but he had come to London to learn new skills, not brush up on old ones. Asking him to draw was like asking someone who is bilingual to translate—no big deal. However, Hewett, who as a young man had hoped to be a painter himself, even studying in Paris, was defi-nitely impressed with Carter's work. He called it "capital," Carter re-ports proudly.

As true in real hospitals as in soap opera ones, there were not a lot of secrets in the corridors of St. George's. Word reached Henry Gray of Carter's abilities, and Gray, it so happened, was in need of an artist's eye. *Would you mind looking at some drawings I've had commis-sioned?* he inquires one day. The discussion that followed must have had an interesting dynamic because, in this area of expertise, Carter was Gray's superior.

The drawings were made for an essay Gray was writing on the spleen. One illustration was "miserable," Carter recalls that evening. Another, "shabby." What's more, in Carter's view, the fee the artist was asking was preposterous. Henry Gray himself, though, made an excellent impression: "Gray very clever and industrious: a good model."

On first reading this entry, I was pleasantly surprised by the nineteen-year-old's forthrightness in assessing the artwork. *Good for you, H.V.* Embedded within this same passage I found something even more illuminating, though it is written so quietly it would be easy to miss: "Offered own assistance." Carter than adds matter-of-factly, "Gray will let do some."

Will let me *do some, he's saying,* I thought to myself, filling in the implied pronoun. At that moment, I felt as though the research gods were smiling down on me. Here was proof that they began working together almost two years earlier than historians have thought, I realized with a start. And it was not Henry Gray who first proposed the idea, as I would have assumed, but H. V. Carter. For him, Friday, June 14, 1850, started and ended like any other day—which is to say, like any other diary entry—but sometime between wake-up and bedtime, a historic partnership was formed.

Four

FIVE DAYS INTO THEIR FIRST COLLABORATION, HENRY VANDYKE Carter could happily report that Henry Gray liked the work he had done, and yet, he admits in his diary, drawing the spleen is "not easy." In fact, the spleen is "not" a lot of things, I am finding, seven weeks into Gross Anatomy. It is not part of the digestive system, for instance, though it's located in the abdominal cavity. It is not part of the urinary system either, though it's connected to the left kidney. It is also not a part of the circulatory system, though its two main jobs are blood-related—recycling worn-out red blood cells and helping produce certain infection-fighting white blood cells. That these cells are called lymphocytes gives a clue to the spleen's actual affiliation: the lymphatic system. The spleen is one last "not," I should add. It is not vital. If it has to be removed due to injury or illness, the body can make do just fine without it.

The spleen is oblong and about half a foot long (fifteen centimeters) and, on the inside, spongy, with two kinds of pulp, red and white, which may have been the aspect Carter found difficult to render. Regardless of its actual appearance, I expect I will always associate the spleen with Denise, the giggly, freckle-faced, Japanese-American student who played the role of the spleen in one of Dhillon's most memorable lectures. The abdominal cavity is frankly a big, twisting, confusing, crowded mess, and he wanted to show us how the parts fit together.

"Imagine we're all sitting inside the abdominal cavity," Dhillon began, sweeping his hands to indicate the entire rounded lecture hall. "See the slide projector up on the back wall there?" Everyone turned in their seats. "That is the belly button." This got some chuckles. "The chalkboard behind me here is the vertebral column, the ceiling's the diaphragm, the floor is the pelvic . . . *what?* Anyone?"

"*Floor,*" someone answered.

"Yes, the pelvic floor, very good, which keeps the poop from falling out." Speaking above the boom of laughter, Dhillon added with faux grandeur, "Now, imagine that I"—he patted his generous belly—"I am the stomach!"

Next, he selected his cast of abdominal viscera and, one by one, positioned each in relation to himself: Gergen, the beefiest guy in the room, became the biggest organ, the liver, just to his right. (Someone gave Gergen a backpack to hold as a gallbladder.) Baby-faced Dan, tall and thin and sweet, became the pancreas and stood directly behind the stomach and next to the spleen, Denise, who peeked out from Dhillon's left side. Right behind everyone, the inseparable gay couple, Andy and Wilson, held hands. They were the kidneys. By this point, the cluster looked like the world's most awkward group hug, but Dhillon had just gotten started. Between the pancreas and the liver, he positioned the aorta (Amy) and vena cava (Ming), then added the ten different sections of the intestines, anus included, even though it's technically not in the abdominal cavity (all the students in the second row were enlisted for this). Finally, he squeezed in the many interconnecting ligaments and membranes that keep all the abdominal structures stable. Spread arms and fingers worked this visual magic.

By the time Dhillon had finished, the cast of twenty-five had become a wriggling mass of simulated digestion. Somewhere in the back right, I could hear the spleen giggling.

Twenty minutes later and twelve floors up, in the lab, a dramatic scene, wardrobe, and mood change has taken place: Massoud removes the thick sheaf of abdominal muscles from our cadaver, and I, standing between him and Miriam, have rarely ever been so repulsed in my life. Lying exposed is a mass of glistening, fat-laden tissue that covers the entire abdominal cavity. This thick membrane is called the greater omentum (or, "large apron"), something I had not previously known existed. In our cadaver, it looks and smells like a rotten jellyfish. What lies underneath is grimmer.

Miriam begins by cutting open the stomach, which is not in the center of the belly, as I would have assumed, but tucked under the lower left ribs. She then reaches in, as if feeling for change at the bottom of a purse, and discovers a bolus of undigested food—meaning, this person had died soon after eating. This is the tipping point for me. My insides churn, the very definition of a visceral reaction, and I excuse myself from the table.

I am standing by the window taking in some fresh air when a squeal from the far corner of the room pulls me in that direction. Stephen and his partners have opened a gallbladder and found gallstones. "Here, Bill, do you want to feel one?" he offers.

Before I know it, I am rolling between my fingers what looks and feels like a black marble. This is *cholesterol*, calcified and mixed with bile pigment. These things can cause a lot of pain when they obstruct the flow of bile from the gallbladder. What's more, people who have gallstones almost certainly have plaque-lined arteries, which is exactly the kind of life-threatening condition that can be treated by prescription drugs, in this case. No wonder Stephen, this aspiring pharmacist, is so fascinated. I tip the gallstone back into Stephen's palm, thanking him for sharing.

To appreciate this dissection, I need to think more like a pharmacist, I tell myself. I need to *see* like a pharmacist. I also have to remember that *being* a pharmacist does not necessarily mean working at a Walgreens, Duane Reade, or Rexall. While some of these young people plan to become traditional community pharmacists, working at a neighborhood drugstore (or, as in the less traditional case of Andy and Wilson, opening their own), many will use their degrees as stepping-stones to larger plans. Stephen wants to go into drug development and join a major pharmaceutical company. Amy, on the other hand, is simultaneously earning a master's in public health and hopes to work for the FDA. Theresa, who already has a master's in criminology, plans to specialize in forensic pharmacology. Her entire focus will be to determine when drugs are a cause of death, whereas

Miriam, who intends to become a clinical pharmacist, will work in a hospital as part of a surgical team. To all of these students, wherever a pharmacy degree may take them, this dissection will form the bedrock of their future work. The alimentary canal is, after all, the route traveled by every pill, tablet, capsule, elixir, syrup, or substance that can be swallowed, and knowing this part of the body backward and forward is essential to understanding how drugs are absorbed, dispersed, metabolized, and eliminated.

Back at my table, Massoud and the others are stuck in the mass of confusion known as the small intestine. Gray describes this part of our anatomy as "a convoluted tube," as if he, too, had gotten lost here once or twice. The group backtracks to the stomach and begins again. It is easy to find the first part of the small intestine, the duodenum, because it is the first ten inches (twenty-five and a half centimeters) off of the stomach. (*Duodenum* comes from the medieval phrase meaning "intestine of twelve fingerbreadths.") Here is where most ulcers occur. The remaining nineteen or so feet (six meters) of the small intestine is composed of, first, the jejunum, followed by the ileum, which eventually leads into the large intestine. Had we tipped our cadaver to the side, the intestines would not have just poured out, and neither could

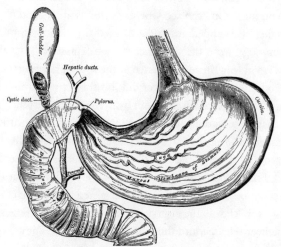

FIG. 492.—The mucous membrane of the stomach and duodenum with the bile-ducts.

we have simply pulled them out, foot by foot, the way you see in gory movies. Connective tissue keeps the twisting mass contained in the abdominal cavity but able to roil freely during the digestive process.

The large intestine, about five feet (one and a half meters) in length, frames the small intestine on three sides and comes in four main sections: the ascending colon up the right shank; the transverse colon across the abdomen at the level of the bottommost rib; and the descending colon down the left, which then turns into the S-shaped sigmoid colon and feeds into the rectum. By any name, bowels are bowels and unbeautiful, but serve a necessary purpose. Within these coils is where water is extracted from what has been digested and what remains is turned into what will be flushed.

"Oh, Bill, did you see this?" Miriam asks, nudging a pinkie-sized flap of tissue that, in fact, I had not noticed.

"The appendix," she says at the same moment I recognize it.

I love it when an anatomical term comes with a built-in mnemonic. True to its name, the appendix is *appended* to a corner of the large intestine. A handbreadth away is an organ that's completely unmissable, the liver, filling the upper right abdominal cavity. Whether viewed in situ or displayed on a specimen tray, the liver is an impressive sight, large and smooth, with perfect lines and a broad surface. It almost looks sculptural. One can see why early philosophers and physicians were fooled by its appearance. Plato, in a dialogue on natural science and cosmology from the fourth century B.C., asserted that the liver played a key role in the maintenance of the soul by keeping the organs of the abdomen in line. The liver did this through its smooth, shiny surface, which reflected images sent from the "divine psyche," the immortal source of rationality housed in the head. Six centuries later, Galen called the liver the seat of life; hence its name, evolving from an ancient root, *leip,* associated with "life" or "living," which later became *lifer* and, finally, the familiar term, *liver.* Galen believed that the liver produced blood—the life force—from digested food, which explained why the liver "clasps" the stomach, he wrote, "as if by fingers." Further, the liver released into the bloodstream Natural Spirits, an incorporeal substance that provided the body's mass.

In reality, this organ filters toxins and medications from the bloodstream; converts blood sugar into usable energy; produces and secretes bile, a substance that helps the intestines digest fats; and performs hundreds of other functions.

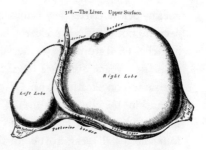

But the liver also possesses a remarkable ability that can compete with the fantastical beliefs of old: alone among the major organs, it can regenerate. In a liver transplant, for example, if you remove half a liver from a healthy donor, the donor's remaining organ will grow back to its former size. And, even more amazing, the half liver transplanted into a recipient will grow to the exact size of the recipient's own original liver. Now, it is impossible to explain *why* the liver would do this. (As Dana would say, "In anatomy, you can't ask *why*. It just *is;* that's how we were made.") But to me, it is hard not to see something of the miraculous in it. We are each meant to *be* a certain way, and our bodies make it so, as if predestination were encoded in our anatomy.

PONDERING HIS FUTURE on the eve of his nineteenth birthday, H. V. Carter summed up his prospects in two sentences: "With energy and perseverance much may be done and without either, nothing can be done. Two roads: to mediocrity and [to] eminence."

Which one would he take?

Mediocre or eminent: which would he *be*?

Though the occasion was special, the dichotomy was typical. To young Mr. Carter, as exacting as he is earnest, everything is black and white—and Gray. Henry Gray starts appearing in the diary almost every day after being promoted to house surgeon in late June 1850, two weeks into the spleen project. Carter goes on daily rounds with him and witnesses his bedside manner, no doubt, as well as how he interacts with medical students and fellow doctors. Less than a week goes by before Carter begins making inquiries about what it takes to become house surgeon; clearly, he has got a goal in sight. He approaches Dr. Tatum, Gray's mentor, who says that it normally takes at least six years. *Six whole years?* Seeking a second opinion, he has the same conversation with Gray, with whom he has a growing friendship. *You* could do it in half that time, Gray tells him, paying the younger man a fine compliment—Gray, after all, had needed four and a half years to obtain the appointment. But in Carter's mind, he had his work cut out for him. Compared to his friend, he saw many failings in his own character. He describes himself as "indecisive," "very slothful," "diffident," "not confident," and, over and over again, "idle." Even after the busiest days, jam-packed with schoolwork, hospital work, and artwork, he reprimands himself. "Must work more." "Must work better." "Must be more exact." *Must,* in fact, must be the most frequently used word in Carter's vocabulary.

So extreme is his self-criticism—and so contrary to the record of his achievements—I actually wonder if Carter had a kind of personality dysmorphic disorder. His diary is the mirror he looks into at the end of each day, and all of his accomplishments appear distorted. However, in his writing, I also hear the voice of a young man who is trying to push himself, to prove himself, to break the mold. A young man who is determined not to be like his father.

Henry Barlow Carter does not make many appearances in his son's diary, but when he does, he leaves an unforgettable impression. One need turn back only a few pages to find a choice example. It is a month earlier, May 24, 1850, and Carter has just learned that his father is going to be "in town" the next day. But the visit gets off to a rocky start. Rather than contacting his son directly with the details of the two-week visit, Mr. Carter instead notifies Dr. Sawyer. "Don't see why Father not write to me," H.V. complains on the eve of Henry Carter Sr.'s arrival.

Carter had lived in London for two and a half years by this point, and although he did not consider himself as worldly as other Londoners, he had matured considerably. Just working at St. George's, its wards crowded with the sick and dying, had been a crash course in growing up. Nevertheless, his father still treats him as if he had never left Scarborough. On their third evening together, for instance, he sits his son down and gives him a "lesson and hints" in art, just as he had throughout H.V.'s childhood. I am sure Mr. Carter felt this was a fine way to spend the time. If his son did not, though, I can't blame him. After all, he no longer considered himself a "student" of drawing; Dr. Hewett and others at St. George's certainly did not think so. Further, he was working in a completely different style from his father.

Even from a vantage point of a century and a half, it is easy to see that the two Carters are headed for a blowout. Two days later comes the damage report: "Have acted foolishly, hastily and improperly. . . . Father much put out. Say hard things." Next comes the silent treatment: Henry Sr. tells Henry Jr. that he does not wish to see him for a few days. By the end of the week, they have made peace, attending church together, but soon thereafter hackles are raised anew. "He, irritable and annoying," Carter scribbles on the night of Wednesday, June 5. "Self, irritated and hasty."

If Henry Barlow Carter had kept a diary, I would bet his entry of the same night would have read something like, "He, sullen and ungrateful. Self, short-tempered and harsh." These two were following a script that's been played out since the dawn of civilization: teenager wants to be seen as an adult; parent thinks teenager has a long way to go. In a bittersweet final scene, our two protagonists meet the night before the father leaves London, and they reach an accord. With that, "Bade good bye," the son writes in his diary.

Whether Carter's father said all the right things or all the wrong things that last night, one thing is certain: the effect on H.V. was galvanizing. Soon after his father's departure, he added a new "must" to his running list: "Must depend on self." Within a forty-eight-hour period, he formed the collaboration with Henry Gray and made two telling moves: He designed an elaborate box to hold his art supplies and ordered "calling cards" for himself. "HENRY VANDYKE CARTER, SAINT GEORGE'S HOSPITAL," they read. With every letter he paid to have printed, it's as though he were underscoring his new identity: *I Am My Own Man*.

No doubt about it, Henry Gray, whom Carter always characterizes in glowing terms—"capital worker," "nice fellow," an "example of industry and perseverance!"—is the kind of person he aspired to be. Yet it is also clear that Carter was following his own distinct path. For instance, he was studying for an apothecary license, a credential Gray never pursued. To modern ears, *apothecary* sounds like a quaint synonym for *pharmacist* or *pharmacy*, it being one of those odd nouns that applies equally to both a person and a place. And, in fact, in England and in much of Europe from the early to the late Middle Ages, an apothecary was exactly that: a druggist who sold drugs in a retail shop, an apothecary. By the time Carter was in medical school, however, this definition was already antiquated. Being an apothecary was actually more like being a modern-day general practitioner, a doctor with a broad knowledge of diagnosing and treating diseases. Whereas Henry Gray was well on his way toward a career in clinical surgery and research, H. V. Carter ultimately hoped to be a country doctor.

By the summer of 1850, Carter had successfully completed two years of medical school. While he had excelled at anatomy, chemistry, and botany, winning top class prizes in all three subjects, one of his prouder achievements had occurred with far less fanfare: "Made *coup d'essaie* at bleeding," he noted on April 9, meaning he had made his first attempt—or, stab—at bloodletting a patient. As he went on to detail, he had divided the procedure into two "operations": cutting into the skin to expose a vein, then slicing it open. "Not a favorable subject," he added, suggesting that the patient was none too thrilled with being Carter's guinea pig. Other days found Carter practicing other necessary skills: "Pulled out tooth of soldier. Tugged at another." Over the next few years, he would have to master minor surgeries such as these, a routine part of an apothecary's practice, and he would need to earn a degree in surgery. Finally, in order to obtain an apothecary license, Carter would have to complete a five-year apprenticeship. He was now more than halfway there.

John Sawyer, the doctor to whom he was "indentured," was a surgeon and an apothecary and something of a holdover from earlier times. Along with his medical practice, he ran an apothecary—excuse me, a *dispensary*, as retail drugstores were now called. While a patient would typically get a written prescription from his apothecary and take it elsewhere to be filled, Dr. Sawyer still offered the traditional one-stop shopping. If you happened to step into his shop at 101 Park Street, you might have found his apothecary apprentice behind the counter. H. V. Carter occasionally filled in as the dispensary's *dispenser*, the person who counted pills and filled prescriptions for laudanum and tincture of belladonna and black draught and so on. Not that he liked it. "Am determined to have as little as possible to do with shop and prescriptions," he once wrote, "which [I] consider as altogether foreign to duty and [a] source of much annoyance." There was, however, one order he always filled without complaint: his mother's.

An apothecary in an apothecary, c. mid-nineteenth century

Eliza Carter is a mysterious figure in her son's diary. She never comes in person to the dispensary and, in fact, has never even visited H.V. in London. She is apparently too ill to leave Scarborough, though what's wrong with her and what drug she needs are not yet spelled out. She exists on the page only as M. "More pills for M.," Carter might jot, meaning he had received a letter from her with a request for medication. In response, he does something that is almost impossible for me to imagine a pharmacy student today doing: he makes the pills himself.

How? He doesn't say. But then, for good or bad, a diarist is beholden only to his own inner narrative. He does not need to explain such things to himself.

Thankfully, the details of nineteenth-century pharmacology are well documented. Pills began as a paste. The dutiful son would have carefully measured his medicinal ingredients, ground them with mortar and pestle to a fine powder, and added a liquid binding agent. Then he would mix. The importance of thorough mixing could not be overstated because each small pill had to have the same potency when the process was completed. Next came the "pill machine," a simple, hinged apparatus that operated like a waffle iron, albeit without the waffle pattern or the heat. The machine pressed and molded the paste into neat rows of pencil-thin pipes. Finally, perhaps after some hardening had occurred, Carter would use a separate instrument to cut the pipes into their proper lengths, thus creating identical-sized pills. These had to dry before he could package them.

To think about each step Carter took is to be reminded that *Rx*, the symbol for prescription, is a Latin abbreviation for *recipe*, a word that, to me, always conjures up home-cooked meals and a mother's love. There is something so poignant about the role reversal at play here, the child being a provider of sorts. No doubt, it gave the young Carter great pride to help his mother, even though he was far from home. Whenever he mentions the pill making, he comes across as efficient and capable—professional. Even so, the intensely personal nature of the task could not have been far from his mind. After all, these were pills destined for his mother's body, and I get the sense that she needed them urgently. Carter moves as quickly as the process allows but, as indicated in his diary, always takes the time to write a letter to his mother to accompany them. The following morning, without fail, the package is on its way to Scarborough.

ONE CANNOT PICTURE the path of a pill through the body without studying the mouth and throat. Which means seeing the cadaver's face, something we had managed to avoid until now. Though I'm not looking forward to this unveiling, I am curious, frankly. Over the past eight weeks, I have constructed a mental image of this frail tiny old woman. I am sure Massoud, Laura, and the others have done the same. Unlike other groups in our lab, though, we have never named our cadaver; that seemed inappropriate somehow. Perhaps my lab partners thought as I did: How can you name someone without looking into their eyes? How can you ever thank them?

Massoud lifts the cheesecloth veil, folds it back over her hair, and takes a step back. Before us is not a face like any we could have imagined. Instead, we are looking at what an anatomist would call the underlying anatomical structure of the anterior aspect of the head. Translation: a face without the skin. Like all of the cadavers in the room, our body had been partially dissected in another class, but none of us had anticipated the extent of that work. The eyes are closed and intact, as are the lips, but the fat that typically pillows in the cheeks has all been cleaned away. What remains is a mask of musculature, forehead to chin, ear to ear. Running over this is a latticework of blood vessels (empty of blood), facial nerves, and lymph ducts, all in faded shades of white. The dissection is flawless, worthy of a fine-ink drawing by H. V. Carter, and, as Gergen is quick to point out, way too good to have been done by a first-year pharmacy student. Our job for the first hour of lab—identifying specific parts of the face, such as the nerves behind a blink and the muscles that make a yawn possible—has been made textbook-easy.

In front of each ear we find a body part that almost begs a description in the form of a riddle: what lies unseen just under the skin of the face and produces a clear, tasteless, odorless fluid? If you answered, salivary gland, you are half right, but if you said, parotid gland, you're dead-on. The parotid, the largest of the three pairs of salivary glands, is surprisingly huge. I lift my fingers to my face just to confirm that I have a pair, and feel a distinct padding over the back curve of my lower jaw. I had always mistaken this for facial fat when, in fact, these are saliva factories. They provide our natural mouthwash, make it possible to lick our lips, and play a major role in the ability to taste. At the same time, they can be defeated by a saltine and, when their function is inhibited by certain medications, can cause dry or "cotton mouth."

To locate the next several items on our lab list, we have to turn to Dana, who arrives at our table with a very large sealed Tupperware container and the sound of sloshing. She first warns us that what lies within may be upsetting to see. Even so, I am nowhere near prepared for what follows. Setting the lid to the side, Dana reaches down with gloved hands and lifts out what I can only describe as a horror: a severed head, split clean down the middle. A human profile from the inside out. I can see between her carefully placed fingers that the face, male, is intact.

"*This* is a hemihead," Dana says, as if making a formal introduction. "Sometimes it is called a sagittal section or a median section . . ."

The visual drowns out her words.

Clear embalming fluid flows down the exposed lobes of the brain, through the nasal labyrinth, down the throat, and over the edge where the neck ends. Dana waits for the last drops to drip off and sets the head face-side-down onto a towel draped against our cadaver's lower legs.

Amy and Gergen physically turn away from the scene, and I think, *Well, that's a mistake; they're gonna have to turn around eventually.* And eventually, they do, but Laura is the first to press in. In my need to focus on something—anything—I land on the brain, with its familiar whorls of gray matter.

"Can anyone identify the root of the tongue?" Dana asks from the distance.

I move down to the mouth area and see what looks like a gargantuan mushroom rooted in the lower jaw. Laura uses a metal probe to point at the base of this mass, and Dana nods.

A tongue in profile does not remotely resemble the one you see in your bathroom mirror; it is far thicker and longer than you would ever expect. Beyond the pebbly portion at the back of the mouth, there is a full third you never see, which curves down into the pharynx, the top of the throat. The tongue is composed of hundreds of taste buds and eight different muscles and is animated by a major cranial nerve. Best known as the organ of taste, the tongue is also a natural contortionist, able to roll and fold, in many cases, and to wag and probe, making it ideally suited for its supporting roles in chewing, mouth cleaning, speaking, and swallowing.

The act of swallowing had taken up a good deal of Dana's earlier classroom lecture. Sitting in the lecture hall, I had tried to record every detail as she spoke, but this was like trying to notate the choreography of a dance while watching it for the first time. I finally set my pen down and tried a new tack, swallowing while I listened and listening while I swallowed, trying to feel each step Dana described. Still, I could not quite visualize a gulp from start to finish.

But with the hemihead to illustrate, I can now see how the various parts all come into play, how, in order to swallow something—say, a pill—the tip of your tongue first presses up against the roof of your mouth, nudging the pill backward. Next, the back of the tongue rises up to force the pill to the back of the mouth. This automatically stimulates nerves that send your soft palate (the fleshy parts at the back roof of the mouth) upward, sealing off the nasal passageway. Now the pill and the water you are swallowing it with won't shoot out your nose. But in order for it to go down your throat, a number of complicated maneuvers still have to occur. As the hemihead shows, the pharynx is a common pathway for both air and food, but the epiglottis, just below the far curve of the tongue, closes off the opening into the windpipe so the pill does not head to the lungs. At virtually the same time, the voice box moves up and the muscles of the pharynx contract, pushing the pill past the epiglottis and—*gulp*—into the esophagus, the muscular tube feeding into the stomach.

By the time the imaginary pill drops into my hypothetical abdomen, a very real but unexpected transformation has taken place: the hemihead has lost much of its gruesomeness—so much so that, when Dana proposes, "Okay, now let's talk about the gag reflex," I think it a fine idea. Which is not to say I suddenly find it pleasant to look at. But, by comparison with our adventures in the abdominal cavity, the hemihead is neat and clean, practically free of fat, and looks carefully packaged. Each part has its own tidy little chamber. It is hard to imagine how a headache could ever fit in there.

After Dana leaves to work with another group, we make our way through all twenty-eight items on the lab list, from the sphenoid sinus (the deepest part of the nasal cavity) to the dangling uvula to the vocal cords. Before going home for the day, though, we have one last bit of business: returning the hemihead to its container. I offer to take care of this while Massoud and the others agree to clean the instruments and rewrap the cadaver.

I gently lift the specimen with both hands. It is heavy, which surprises me. A typical human head weighs about twelve pounds; this feels closer to a twenty-pound dumbbell. To protect the hemihead for storage, I need to bundle it in gauze, but first I turn it just enough so that I can look at the face. A pale bushy eyebrow perches above the closed eye. The

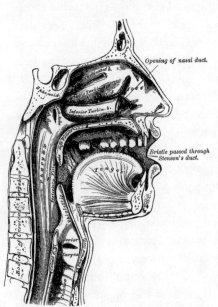

FIG. 466.—Sectional view of the nose, mouth, pharynx, etc.

nose must have been impressive in its entirety. *This was someone,* I think, caught in an upswell of awe, *a thinking, dreaming person.* Judging by his wrinkled skin, I would guess he was around eighty at the time of his death. Who knows if he lived a good life or not? Maybe he had been a criminal, maybe a doctor. Maybe a criminal doctor. I would never know, and, in truth, that was neither here nor there. What mattered was that he had donated his body so that others could learn.

I make quick work of the gauze, then carefully lower the head into its bath of embalming fluid. The lid makes a *thoomp* sound as I press it closed.

Five

ETWEEN THE BOTH OF THEM, THE TWO HENRYS WON JUST about every academic honor offered at St. George's during their years in medical school. For the year 1850, for instance, Carter received the annual Botany Prize—meaning he was his class's top-ranked student in the subject—as well as smaller prizes in three other subjects, including, no surprise, anatomy. The awards were distributed during the opening ceremony for the school's winter session.

In Carter's diary entry for that October day, he sounds less than thrilled about having had to attend the "affair." This was his third year in a row as a prizewinner, plus, even though classes had not yet begun, he was already deeply immersed in work. He had just started a new collaboration with Gray, an investigation into the development of chicken embryos, which involved much cracking of eggs, and he was also busy with his own studies. In fact, attending the ceremony meant setting aside a fascinating dissection at Kinnerton Street, changing into dressier clothes, and trudging over to St. George's. Nevertheless, Carter reports being quite impressed with the closing speech given by Benjamin Brodie, who presided over the event. The revered doctor "excited ambition and," Carter adds cryptically, "gave warnings."

Curious about what Brodie might have said that so roused the nineteen-year old, I make another visit to the Special Collections Room at UCSF. The library houses a first edition of *The Works of Sir Benjamin Collins Brodie*, a rare three-volume set from 1865, and I am hoping it will contain some mention of Brodie's long-ago speech.

I arrive to find that Ms. Wheat has already retrieved the books I'd requested and placed them, along with a clean pair of white gloves, on the broad reading table. As on past visits, I have the room to myself. I am prepared to spend hours searching through the thick volumes, all eighteen hundred pages, but no sooner have I taken the first from the stack and turned to the table of contents that I find exactly what I am looking for, even listed so that I cannot possibly miss it: "Address on Delivering the Prizes to Pupils of St. George's Hospital."

"*Wow,*" I exclaim in my loudest quiet voice. Ms. Wheat, tapping at her computer, shoots a smile in my direction.

Turning to page 532 almost makes me a believer in psychometry, the ability to pick up impressions from an object simply by touching it. I begin reading Brodie's speech and can picture the whole scene: the crowded auditorium, the stage where Carter and his fellow prizewinners are seated, and Dr. Brodie at the podium, with his Ichabod Crane face and wavy mane of gray hair. He addresses the student body collectively. "First, let me impress on your minds that the next few years are the most important and critical period of your lives. You are now to lay the foundation of that knowledge on which your future character—nay, your very subsistence—is to depend. Let these years be wasted, and you will never be able to redeem the loss. Ceaseless but unavailing regrets will haunt you during the remainder of your days—"

Yeesh. There are those "warnings" Carter mentioned.

Soon, though, Dr. Brodie's tone softens. "I leave it to your respective teachers to tell you what lectures to attend, what time to devote to the dissecting-room and hospital," and so forth. But he had one sage thought to add: Get in

Sir Benjamin Collins Brodie

the habit of taking clear, careful notes of all lectures and cases. "The notes thus taken should [then] be transcribed in the evening, and preserved for future use. You will find them the best things to refer to—as far as they go, much better than books—in after life," by which he means life after graduation. So important does he consider the practice, Dr. Brodie tells the audience, he awards his own yearly prize to the St. George's student with the "best series of clinical notes."

Judging simply on penmanship, Carter would never have stood a chance of winning this particular honor, in my opinion. Still, I find the idea of a Note-Taking Prize enormously appealing. Were UCSF to resurrect it, I have no doubt the winner would be Ming, a pharmacy student I've gotten to know over the past couple of months. Her notes make mine look feeble, although, granted, our styles differ greatly. During the minilectures in lab, I simply jot words and phrases onto a small pad that fits into my scrubs shirt pocket while Ming records sentence after sentence on sheets of graph paper in tiny, perfect print. She uses a four-color Bic pen, the kind I have not seen since junior high—red ink for notes on blood vessels, black for nerves, green for muscles, blue for organs—clicking from one to another with barely a glance up.

It was over the topic of note taking, in fact, that Ming and I bonded during one of the first labs. I happened to be standing next to her and saw her in action.

"Those notes are beautiful," I said in all sincerity in a pause between her clicks.

"Oh, these are just rough," she replied, and not out of false modesty. Ming planned to rewrite them once she got home, combining her lecture and lab notes and supplementing them with snippets from the textbook. She would then transfer those to a three-ring binder, color-coded by course subject. Now that's my kind of obsessiveness.

From this meeting on, I would regularly drop by her table and say hello. Ming had a sense of style as quirky as her personality— bohemian chic meets Tokyo pop. She loved hunting for vintage clothes on nearby Haight Street. Up in the lab, though, beneath over-sized goggles, rubber gloves, and a disposable full-length smock (thrown out after each lab), she looked like a holdover from the film *Outbreak*. She had made it her job to take notes, holding her Hello Kitty clipboard like a shield in front of her chest. This left her no time to actually participate in dissections. But, as I came to under-stand, this was fine by her. The copious note taking was, intention-ally or not, an avoidance strategy.

Sometimes I would spot Ming at the opposite side of the lab, standing at such a remove from her group that I'd feel the urge to walk over and give her a gentle push from behind, just so she would actually be in the huddle. And once she took that actual step, I felt a genuine sense of brotherly affection for her. Late one afternoon, after most of our fellow students had left, I pulled Ming over to my group's cadaver. I suggested we review the vessels of the neck to-gether. At one point, she got so caught up that she set her clipboard down and, without really thinking, used her prize pen to poke at the artery we had been searching for.

"I cannot believe I just did that!" she said, laughing and brushing it off on her smock. On second thought, she went over to the sink and washed it. Returning to the table, Ming picked up a metal probe with a tiny flourish and we continued studying. The following week, she told me with pride, she performed a dissection by herself.

Several weeks later, the day before my visit to the library, I ran into Ming as I was heading up to the lab. This would be the last lab before the final exam, so it seemed natural to ask how she thought she had fared in the course.

A sheepish look washed across her face. "Well, I kind of had a hard time at first."

Yeah, I conceded, I kind of noticed.

While I'd assumed that Ming planned on working at a Walgreens or someplace similar, she confided that, in fact, she hoped to be a clinical pharmacist in a hospital. Deep down, though, she hadn't been sure she was cut out for it. "This may sound really weird, Bill, but working with the cadavers has actually made me feel more comfortable about working with patients, if that makes any sense." She smiled to herself. Then, in a manner that told me she absolutely will, Ming said, "I think I'm going to do fine."

"I do, too."

I had come by the lab mainly to say some goodbyes, since I would not be coming to the students' final exam. I chatted with Massoud and Laura but didn't see most of the others I had gotten to know. This was an optional review session, and on this beautiful afternoon, many had exercised the option to skip it, understandably. I, too, decided to cut my stay short. I waved to Dana and Sexton, who were bouncing back and forth among three small groups—I would catch up with them sometime later—and went to retrieve my bag from the windowsill. A couple of steps away, Dhillon was supervising two students I did not recognize. "What are you guys working on?" I asked.

One of the two explained that they were third-year med students practicing surgical techniques. He told me wryly he was trying his best "not to make too much of a mess" of the head. From where I stood, it did not look as if he was succeeding. As for his friend, he was a slight young man holding an electric bone saw above the chest of a massive cadaver, aiming for the sternum. He was the very picture of apprehension as he turned on the saw, which I took as my cue to move on.

At the far end of the lab, I saw another curiosity: seven people in street clothes sitting on stools in a semicircle, holding sketchpads in their laps. They were from the Art Institute of California. Once a quarter, they come here to sketch as an adjunct to their life-drawing class. Laid out on the table before them was not a cadaver but a selection of expertly dissected specimens—prosections, they are called (a shortening of the term *professional dissections*)—a still life of preserved arms and legs, a torso and a head, their muscles and vessels bared.

"It really helps to be able to see what's under the bumps," a young woman told me in all seriousness.

I couldn't help smiling: *bumps.* I liked that.

I asked the student nearest me if I could see his sketches. He was a big pasty-skinned guy, the kind one would guess spends an inordinate amount of time playing video games. And, in fact, he told me that he is studying character modeling at school, with the hope of becoming a video game designer. He flipped back a couple of pages on his pad so I could take a look. His sketches were good—not H. V. Carter good, mind you, but they had dimension. In one drawing, he had reassembled all the separate parts before him into a full anatomical figure.

I complimented him, but he shook his head and looked up from his pad with doleful eyes. "Drawing at the zoo was a lot easier," he said.

The UCSF staff person overseeing the art students' visit was a woman named Andy, whom I had seen many times but never met till now. Andy explained that she does everything from ordering the cadavers and lab supplies to scheduling classes and cleaning up. Just then, the lab's wall phone began ringing. "Oh, and I answer the phone, too," Andy said as she rushed to get it.

After a minute, she rejoined me. I told Andy that I had been attending Dana's class and had learned more about anatomy than I ever could have imagined. "I'm actually going to kind of miss coming to the lab," I admitted.

Andy nodded, as if she knew exactly what I meant. Her eyes sparkled behind small, thick rimless lenses.

Before leaving the room, I clutched the doorframe and took a last look around. Even with the harsh lights, the well-scrubbed linoleum, and the funk of formaldehyde, it seemed less like a lab than like a library—a place where not only human anatomy but the spirit of anatomical discovery is preserved. And there, in the far back corner, I could easily imagine a small man in a black coat, Henry Gray. He had been here all this time, silently working.

BACK ACROSS PARNASSUS STREET, inside the Special Collections Room, Dr. Brodie is bringing his speech to a close. He turns to H. V.

Carter and his fellow honorees: "To you, Gentlemen, who have been the successful competitors for the prizes given annually by the different teachers of this school, I offer my sincere congratulations. If you have gained honour for yourselves, you have also done good to others, for example is better than precept; and there is no one among you who has not exercised a wholesome influence on his fellow students."

At this, I can imagine applause breaking out in the St. George's auditorium.

Dr. Brodie waits for silence to return. He then adds, in a voice so resonant, so full of wisdom, that it carries across the centuries and reaches me here: "Let me advise you to pursue the same course through life, recollecting that, even as practitioners, you must still be students. Knowledge is endless, and the most experienced person will find that he has still much to learn."

Six

*T*WELVE DAYS LATER, I AM BACK IN THE ANATOMY LAB. IT IS DAY one, number two—the first lab of a new course—and virtually everything is different: the time (9:00 A.M.), the teachers (no more Dana, Dhillon, or Sexton), and the class size (just twenty-six). Even the students themselves look noticeably different—more athletic, more tactile and engaged with the physical, which is fitting as these are physical therapy students. But the most striking change of all is with the cadavers. They are fresh. In fact, they are as fresh as medical school cadavers can be—only six months dead—which makes them ideal for the coursework ahead. The focus of the class is neuromuscular anatomy, how the body moves and how sensation—pain, in particular—is transmitted and felt. With the permission of course director Dr. Kim Topp, I will be attending the thrice-weekly hourlong lectures as well as the three-hour labs. Dana had put in a good word for me.

This is not only the first lab of the session but also the students' very first day in the UCSF master's program in physical therapy, so they have barely had a chance to meet one another. Nevertheless, I, in my green scrubs, would never be mistaken for one of them. Dr. Topp requires her "PT" students to wear white lab coats, which make them look more like junior pharmacists than the pharmacy students ever did. Every so often, another dress code will be enforced, I have been told: sports bras and shorts, to be worn on days when the class will be supplementing the study of the dead with the anatomy of the living—themselves.

"So, do we just start?" murmurs Kristen, one of the four students at the table I have joined.

For several minutes now, the entire class has been silently standing at their assigned tables, waiting for instructions from Dr. Topp, who is slowly walking the perimeter. None come.

Another minute passes before I see the message sinking in: the class is expected to have read the syllabus in advance and to get right to work. "Yes, I think so," I whisper back to Kristen. "Here, let me show you how to put a blade on a scalpel."

The others at the table are Kelly, Cheyenne, and one of the few males in the class, Sam. The sixth member of our group is a sixty-two-year-old female who, according to the "Cause of Death" list on the side wall, had died of a stroke. The only incisions on her body are the small cuts at the neck and inner thigh where the mortician had injected the embalming fluid into major blood vessels, thereby using the circulatory system for one last go-round. After a half year spent in darkness as the preservatives preserved, this morning marks the body's return to light.

The skin feels moist and supple and, though cold, surprisingly life-like, thanks to a wetting solution whose key ingredient is, believe it or not, Downy fabric softener. Surrounding the cadaver are puddles of clear liquid, as if it had been sweating profusely in its vinyl body bag. By smell alone, I recognize this as excess embalming fluid, though the fumes do not bother me as much as they once did. There is no scent of decomposition.

The head is wrapped in gauze and covered in clear plastic, as are the limbs, which are tied together at the ankles and wrists. Kristen snips the twine in both spots, then she and I each pull an arm to the side. "Okay, that's better," Kristen says to herself, and she's right, it is. Tied up, the cadaver had looked like a kidnap victim for whom the ransom arrived too late.

Kelly returns to the table with a rectangle of thick gauze, which, in a respectful gesture, she places over the cadaver's genitals. Sam begins to read aloud from the lab guide. The main assignment for the week is to study the muscles of the upper body, but, first things first, we have to strip the torso of its skin and underlying fat. Step one: remove the skin from the chest.

Taking the lead, Kristen, positioned on the right side of the body, makes a crosswise incision from the top of the sternum to the small bump of bone where the shoulder blade meets the collarbone, the acromion process. (Every one of our body's "bumps," as the art student had referred to them—or *processes,* as they are technically called—has a name.) Without a word, she hands the scalpel over to Sam, who makes an identical cut on the left side. Sam passes it to Kelly. She slices straight down the middle of the chest to the naval, and Cheyenne continues with a cut to the outer right side of the abdomen. Next is my turn, which, for me personally, is a major milestone. This will be my first incision.

Wanting to get a better feel for how Gray and Carter had jointly performed the dissections for *Gray's Anatomy,* I had asked Dr. Topp earlier if I might participate, not just observe, during lab. "Well, *that's* a first," she had responded with a warm laugh, noting that she typically gets the opposite request, students asking to be *excused* from dissecting. She said I was welcome to take part but reminded me that the PT students would be graded on their dissecting skills. I promised not to ruin anyone's GPA.

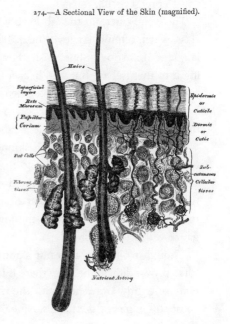

274.—A Sectional View of the Skin (magnified).

The blade's tip sinks easily into the skin and meets no resistance as I pull it across the cadaver's abdomen; it feels as if I am slicing a soft piece of leather. I'd feared that I might cut too deep and damage underlying muscle, but in my cautiousness, I only give the cadaver the equivalent of an eight-inch paper cut. I retrace my line, then set down the scalpel.

Though it is only ten after nine, Kristen, Kelly, Cheyenne, Sam, and I have already become a team.

With the double doors incised, we now have to pry them open, then off. Working in rotating pairs on each side of the body, one person grasps a tiny corner of skin with forceps and lifts while a partner slices the underlying fascia—connective tissue—with a scalpel. To do it cleanly is slow, painstaking work. When on scalpel duty, I find that using a gentle sweeping motion, as if wielding a miniature scythe, works best. Another helpful tip comes courtesy of the lab guide: after you have peeled back several inches, make a *buttonhole* in the skin—yes, that's the word used—and hook your finger through it; this allows a better grip than with forceps. One thing I will not soon forget is the sound of skin being pulled off, a tearing sound, like old Con-Tact paper being torn from a shelf. With the removal of the skin, what remains on the torso is a lumpy coating of bright yellow fat. It does not just scoop off; it has to be either cut or plucked away with tweezers. Along with the fat, we carefully remove all of the cadaver's breast tissue.

The atmosphere in the lab is very different from the pharmacy class, which often had a jovial buzz. There is no chattering here, no laughter. If anyone is squeamish, they are doing a good job of keeping it to themselves. But I do hear lots of sniffles, and at my table, everyone is teary-eyed, as if listening to Sarah McLachlan on an endless loop. This is the classic reaction to formaldehyde fumes. The farther we get into the cadaver, the harsher the smell.

Once the front of the torso is skinned and defatted, the musculature is exposed from the collar to the waist. We now must do the same to the back. The body is a good 170 pounds and the very definition of deadweight. Plus, it's wet. So turning it over takes all five of us. We place a wood block at the forehead so the nose under the gauze will not be crushed.

Though we have gotten our skinning technique down, the rear torso proves difficult because there is less subdermal fat and the skin is more tightly enmeshed with the muscles of the back. The skin comes off in pieces, not panels. Ninety minutes later, though, our cadaver shows an impressive-looking vest of musculature, and all we have left to do is clean up our table. But this raises a somewhat uncomfortable question. As Kelly bluntly puts it, gesturing to the pieces piled on paper towels, "What do we do with all this skin?"

The short answer is, we throw it away, and I cannot help recalling how troubling this had been for one of the pharmacy students, a young man named Edo. After the second lab, as he and I left the building together, Edo had something of a meltdown. "We're just tossing out parts of a human being," he said. "Into a wastebasket!" He seemed in shock at what he had been party to. "It just feels wrong. That's going to be *us* someday." I understood what he meant. And yet, as I told him, I think there's another way to look at it. Aren't we treating the body with respect by using it for its intended purpose—to learn—and shouldn't properly disposing of the remains be part of the ritual?

I point out to Kelly the large red medical waste receptacles placed throughout the lab. "We put it all in those," I tell her and the others, "—skin, fat, anything from the body." The material is saved and later cremated along with the cadaver. At the end of the academic year, the school holds a memorial service in tribute to those who gave their bodies to the program.

Kelly nods thoughtfully. "I'll take care of it," she says. The rest of us divvy up the remaining housekeeping duties, and, soon after noon, day one comes to a close.

The physical therapy class is small enough that, by the second lab the next morning, I've had a chance to meet all of the students. I am surprised by the range of specialties my fellow dissectors are considering once they have earned their PT degrees. Some, including Sam, plan to go into sports medicine, helping athletes get back in the game after career-threatening injuries, while another few, Cheyenne among them, are interested in working with individuals who require a much longer and more gradual course of rehabilitative therapy, such as victims of stroke or spinal cord injury. Kelly and a few other students plan to specialize in postsurgical PT, helping patients recover from mastectomies, for example, in which the loss of tissue and muscle requires relearning basic movements. One woman told me she hopes to work with pediatric patients, another with the elderly.

Whatever their chosen field, working with muscles is in all of their futures—and the body has some 650 to study. Today, we have to identify just 15, the major muscles of the back and shoulder, which, in the abstract, sounds quick and simple, but in reality, not so much. What makes it challenging is that musculature in an actual body rarely looks like musculature in an anatomical illustration, where each muscle is helpfully differentiated with shading and texturing. In the body, individual muscles often lie or fold so closely together that it is hard to tell where one stops and another begins. To identify them, we will need to find what look like seams between the muscles—"fascial planes," these are called—and tease them apart with our fingers.

Our group is luckier than most, for our cadaver has an unmissable trapezius, the diamond-shaped muscle extending from the very top of the neck to the middle of the back and from shoulder to shoulder. It is so large and well developed, in fact, that the five of us wonder aloud what this woman could have done for a living.

"Probably a nurse," offers Kristen, "you know, having to lift patients—"

"No, I'll bet she was a truck driver," muses Cheyenne in her *Fargo* accent.

"Yeah, a truck driver," Kelly agrees, "and that might explain the stroke: too much fast food." This gets nods all around, though I am still wondering why no one has suggested *physical therapist*.

Using the trapezius as our compass (its attachment at the occipital bone being north), we are able to get oriented. Directly to the east and west, we find the deltoid muscles, which look like epaulets draping each shoulder. Deep to the deltoids are the four rotator cuff muscles and, just medial to them, the major and minor *rhomboideus* muscles, so named for their tilted rectangle, or rhomboid, shape. I have known of these muscles for years but only in the parlance of the gym—*delts* and *traps* and so forth. When seen as flesh, they command a higher level of respect, it seems to me. It just would not sound right to use *lats,* for instance, to refer to the huge sweep of muscle rising from the small of the back to the axillae, the armpits. (This is the muscle—a *single* muscle, in fact—that gives bodybuilders that distinctive triangular silhouette.) No, here in the anatomy lab, it deserves to be called by its full formal name, the latissimus dorsi.

The musculature of the back goes four layers deep. Leaving one side of our cadaver intact, we carefully dissect the other, exposing the intermediate and then the deep back muscles. Within an hour, we have reached the very core of the body, the long, vertical erector spinae muscles lying in grooves on each side of the spinal column. These are the primary "good posture" muscles, vital to keeping our

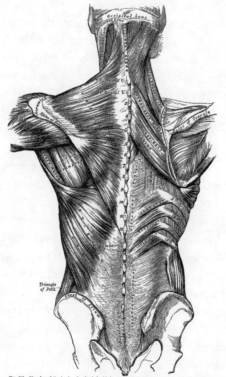

FIG. 213.—Muscles of the back. On the left side is exposed the first layer; on the right side, the second layer and part of the third.

backs erect, and, indeed, they do look powerful. I pull out my notepad to jot down my observations, though my impulse is not to describe the muscles but to draw them, to rough out a map of the back. In truth, my sketch looks like something scribbled in the middle of the night while half-asleep—a doodle from a dream. But when I get home and check it against my own back in the bathroom mirror—shirt off, sketch in one hand, hand mirror in the other—I can see the same shapes under my skin.

Wonderful, I think. *I have the back of a sixty-two-year-old female truck driver.*

I feel faintly ridiculous doing this but still find it cool. I angle for a look at my lower back and, in the act of turning the mirror, turn back thirty years. Clear as can be, I see myself at fifteen in my bedroom in the basement of my childhood home in Spokane: using the blue plastic hand mirror I've borrowed from my sisters' bathroom, I am trying to get a look at my lats in the full-length mirror on the back of my door. Scattered on the floor in front of me are the following items: a set of free weights, a recent Christmas present from my parents; the *Universal Bodybuilding* manual that came with it; a cloth measuring tape lifted from Mom's sewing table; and my journal, which, since starting my weight-lifting regime, doubles as a training diary.

I had discovered weight lifting just five months earlier in freshman PE. To my own surprise, I'd found that I was naturally pretty strong, more so than most of the boys, so I had kept at it. This was probably the closest I had ever come to being good at a sport. Though I was not a "ninety-pound weakling," I also wasn't tall or self-confident or immune to being picked on. Making myself bigger would be a way to repel certain boys and, had I been truly honest with myself at the time, maybe a way to attract certain others.

I could already tell I was making gains with the weights. The proof was right there in my journal, where, along with my written entries, I had been charting my "stats"—measurements of my chest size, biceps, calves, even my neck. And it still is there, I find after digging up those old spiral notebooks, each one dated and signed upon completion. Though I hadn't looked at any of my journals in a good five years, my response had not changed since the last time: I would be mortified if someone were to read these. *Burn 'em, for God's sake,* I tell myself, *the whole box, right now.* But I know I could not strike the match. Like H. V. Carter, who lived to old age yet never parted with his diaries, I expect I will do the same.

To someone who has never kept one, this may be the hardest part to understand: why save a source of embarrassment? And for the diarist, this may be tough to explain. The attachment is not entirely logical; nor is sentimentality alone the motive. You come to anthropomorphize this extension of yourself. However raw the day, the diary absorbs every word, every ache or joy, its blank pages inviting ever more confession. Whether it is a gilded leather-bound volume or a simple file on your laptop, the idea of destroying the diary becomes increasingly unthinkable. It would be like throwing away pieces of flesh from your *own* body. Still, that's only half the explanation. The truth is, when you're writing a diary, a part of you hopes it will be read someday. At the very least, you are writing for that unique someone who will be the perfect reader, who will devour your sentences and *understand:* your future self.

When the nineteen-year-old Carter writes on January 3, 1851, that he had gotten "shaved" that day for the very first time, it seems like

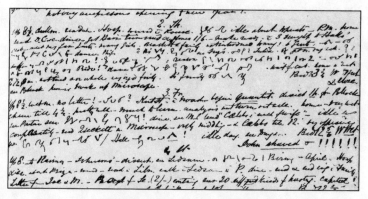

The diary of H. V. Carter, January 1851

something I would have done, right down to the six exclamation marks he uses to punctuate the announcement. Though diaries do not come with rules, all diarists know to record such events, the experience of the new, the starting points. Some of Carter's firsts are unfortunate, as when he falls asleep during a lecture and, even worse, snores. ("Never did so in life before! Humiliating!") Some firsts are nerve-racking but exciting, such as delivering his first speech before fellow students or being called out on his "first midwifery case." ("Labour very easy, 3 hours, disturbed not.") Some are sweet, such as his first secret rendezvous with a girl. And some find him a witness to history, as at the opening of the Great Exhibition, the first world's fair, presided over by Prince Albert and held under the soaring glass arches of the Crystal Palace, not far from Kinnerton Street.

In the life of every diarist, the first of the year is a high holy day, a time for reflecting, for resolving, and, inevitably, for renewing your commitment to your diary. As expected, H. V. Carter—overachiever and micromanager that he was—goes to greater lengths than most would on this occasion. He completes a month-by-month breakdown of the year past, the highs and the lows, cross-referenced to the relevant day, as well as providing his typical summation of the day itself. On Wednesday, January 1, 1851, for instance, he notes that he stopped by the hospital in the morning, read several chapters of the recently serialized *David Copperfield* in the afternoon, went to church, then read some more, and that was pretty much it. "Not very auspicious opening of New Year!" he signs off, as if yawning at his lackadaisical day.

But no, that was not the complete story. In fact, on this day, Carter had begun keeping a secret from his keeper of secrets, a discovery I made quite by chance.

UPON TEARING OPEN the large envelope from the Wellcome Library in London, my first thought was, *Oh no, they made a mistake.* The document I ordered had clearly been reduced during the photocopying.

How had they not noticed? H. V. Carter's handwriting was almost too small to read. And on closer inspection, I could make out just enough to wonder if they had sent me the wrong document altogether.

Manuscript number 5819 was described in the archive inventory as a record of Carter's thoughts on his "religious life as a Dissenter"— his allegiance, that is, to a church other than the Anglican Church, the official Church of England. The document was just fifty loose pages, so I had requested photocopies rather than microfilm. What I held in my hands, though, was not the religious tract I had expected. Rather, it was a diary. Another diary. A shadow diary, of sorts, which Carter had started just as the clock struck midnight on January 1, 1851.

He didn't call it a diary, however. He christened it *Reflections,* a name meant to evoke the more meditative and philosophical bent of this endeavor. But the name would prove to be even more apt than he could have realized.

Here, in a startling change from his usual writing voice, Carter addresses himself in the second person, as if he were looking in a mirror and his reflection yells back.

"You have capabilities which you may reasonably suppose are rather above ordinary," the diary begins. "And how have you used them?" He then excoriates himself for having failed to keep his "invincible" resolutions of the previous year as well as for his immoral behavior—having a foul mouth, habitually lying, losing his temper. Carter is harder on himself than usual but also much more candid. "Your mind," he writes, "[has been] polluted with constant visions of a sensual character," especially at night. And this "loose kind of flirting" you're engaging in with John Sawyer's daughter Mary is "unworthy of a student." You are "deceiving yourself and her, and her parents," and, worst of all, have "no proper end in view."

"Really? Mary Sawyer?" I murmured when first reading this. *I thought you two just played chess together.*

In spite of his suffering, I was euphoric. I felt as if my rapport with this man from another century had suddenly transformed, deepened. I had earned his trust, and now he was letting down his guard, completely.

Truth is, this had not just happened over the course of a few pages. By this point, I had already logged countless hours poring over hundreds of Carter's daily entries. With Steve's code-breaking help, I'd figured out when H.V.'s *H*s meant Hewett or Hawkins or Harland; that he used the German character β to indicate a double *s;* and that nearly every written word ending in *y* ran right into the next word, without a break. I had gone with H.V. the first morning he swam in the Serpentine, the winding artificial lake in nearby Hyde Park, and on every single dip thereafter. I had ridden with the young doctor in the brougham to his first case as surgeon's assistant. He had hardly said a word then. And I had endured, as had H.V., many a "dull m." (dull morning). But at last, all that time spent in front of the microfilm projector at the library was paying off.

In one respect, though, I realized I had been misled. I'd gotten used to seeing his diary pages filling the view screen when, in fact, the photocopies showed both the true size of both diaries—a mere 18½ by 11½ centimeters, or about 7 by 4½ inches—and the scale of his handwriting. My word, the man had an ant's penmanship! He could fit fifty lines on a single page.

Carter added to *Reflections* about every two weeks and would do so for the next four years. Unlike with his daily diary, though, these entries were never meant as an exercise in self-discipline. Rather, like the pages themselves, he was unbound here, writing long, ruminative passages that often read like memoirs. Religion definitely drives this narrative, but he never names his denomination (though it is clearly Christian), rises up to defend its tenets, or, for that matter, lashes out at the powerful Church of England. No, the lashing is always self-inflicted. In *Reflections,* Carter chronicles his efforts to reconcile his moral failings with his desire to lead a strict Christian life. While he continued to chart in his daily diary such details as his attendance at church (sometimes even three times on a Sunday), here he dug deeper, confiding about the battle raging within him between "sensuality—that great bane" and "religion—that important subject."

By "religion" he really meant faith, an unshakable belief in God.

But he really did treat faith as if it were a subject, a skill to acquire. He pursued it doggedly, as if competing for another academic prize. And this one, he *really* wanted to win. "'Tis just the same as with your ordinary studies," he tells himself, "only even more perseverance is wanted." At church, he would take notes during sermons, then write them out fully that evening. He would pore over the Christian tracts his mother sent him and reflect on her great purity of heart. Further, he took up a serious study of the Bible, "comparing texts with a view to getting <u>precise</u> knowledge." And herein lay a crucial flaw in Carter's efforts: he was trying to *know* his way to faith, a feat no more possible than thinking your way to love.

The harder Carter tried to feel God's presence, the more elusive God seemed to become. One evening in July 1851, he writes, "You had reason to think the Holy Spirit had roused you, but alas, an amazing and fearful backsliding has occurred." In the twelve days leading up to this confession, he had tallied an entire "column of filthy sins," while at the same time, all "thoughts of God" were "absent." And for this, he entirely blamed himself.

Carter did have faith in some things—faith that faith existed, faith that other people had faith in God—but he had at least as many doubts. Nowhere is this more apparent than in an intricate chart he created in *Reflections*. Here, he put his internal debate down on paper, weighing the pros and cons of being a devout Christian. To begin, he listed thirteen arguments in favor of choosing such a life and then, in a corresponding column to the right, an equal number against, each one effectively canceling the other out. For instance, number 8: he would have "general serenity of mind" as a Christian and, yet, "frequent mental conflicts."

On first glance, I was amused by this elaborate entry, dated October 1851 and spilling over onto two full pages. The rambling title scrawled atop the chart made me smile: "In Deciding for a Christian Life in Future: A comparative statement of arguments For and Against, Advantages and Disadvantages, Encouragements and Discouragements, in a worldly point of view."

"Worldly" it was not, to my eyes. On the contrary, the chart

looked like the work of an endearingly naïve young man. It reminded me of a similar chart in Carter's daily diary that served as a
ledger of correspondence, letters received scrupulously balanced
against each one sent out. But the more I studied this entry, the more
heartbreaking it became. Along with his concern about whether he
would ever be a good Christian was a second painful issue, though
he intertwined the two: the repercussions of being a Dissenter. The
term *Dissenter* applied to any non-Anglican denomination, but certain religions at the time were far more marginalized and despised
than others, such as Evangelicalism, considered the most extreme,
conservative branch of Christianity. Whatever his specific faith,
Carter definitely saw himself as part of this minority, or, in other
words, as a religious outsider. If publicly identified as such, he would
suffer "jeers and ridicule," "persecution—open and concealed,"
"constant humiliation," "no sympathy with many," and on a personal level, "depression."

As I read of his fears, I felt great sympathy; moreover, I found myself identifying with this tortured Christian English Victorian diarist,
a man both 25 years younger and 155 years older than I am right now.
The question he was agonizing over, boiled down to its essence, was
the same question that had plagued me as a young man: should I
come out? In his case, should I come out as a Dissenter? Pausing over
these two pages, I wished for him what I myself had desperately
wanted at age twenty: for that perfect, knowing someone to show
up and say that perfect, knowing thing—the answer to everything.
Absent that, I was glad that, like me, Carter had a place to pour out
his soul.

Compounding his isolation, Carter was unable to speak of his
struggle with friends such as Henry Gray. He feared jeopardizing his
standing at the hospital. (Henry Gray, as the son of an employee of
the royal family, was presumably a loyal son of the Church of England.) Nor did he confide to John Sawyer and his family. (Though
Dr. Sawyer was not religious, his wife and daughters were Dis-
senters, though, as Carter implies, of a diametrically opposed de-

nomination.) And to his fellow parishioners, he did not dare voice his doubts, lest he be labeled a skeptic.

While it doesn't lessen the poignancy of H. V. Carter's situation, with hindsight one can see that his intensely private struggle was a reflection of larger conflicts arising during the mid-Victorian era. Increasingly, religious belief was being challenged by the new science, Charles Darwin's emerging theory of evolution particularly. At the same time, however, the faithful were in some cases turning *to* science—and to scientists—searching for ways to reconcile science and religion. H. V. Carter was a particularly avid reader of the work of William Paley (1743–1805), for instance. Paley, an English scientist and theologian, claimed that proof of the existence of God could be found in our physical bodies.

Though Paley wrote chiefly in the late eighteenth century, his words resonated well into the nineteenth and, in fact, can still be heard today in the "antievolution" movement advocating intelligent design. With Paley, it is important to bear in mind, however, that he was writing before Charles Darwin was even born, not in reaction to Darwin's theories.

In his most popular and influential work, *Natural Theology; or, Evidences of the Existence and Attributes of the Deity*, Paley introduced his now-famous metaphor of the watchmaker: if you had never in your life seen a watch, then found one lying on the ground and examined it, you would come to the inevitable conclusion, Paley writes, "that the watch must have had a maker—that there must have existed, at some time and some place or other, an artificer or artificers who formed it." One must come to the same conclusion, he argues, regarding far more complex structures—plants, animals, and human beings. In short, only an "intelligent Designer" could have created them, just as only an "intelligent watchmaker" can make a watch. "That Designer must have been a person. That person is God."

While nature offers multiple proofs for the existence of God, Paley singles out one as definitive: "For my part, I take my stand in human anatomy." In making his case, he moves beyond the biblical teaching that humans are made in God's image. Paley presents what amounts to an anatomical travelogue of faith, pointing out those parts of the human body so perfect in design and purpose that God's hand is obvious: "The pivot upon which the head turns, the ligament within the socket of the hip-joint, the pulley or trochlear muscles of the eye, . . . the knitting of the intestines to the mesentery," and on and on, as if he somehow sees each piece of evidence right through the skin.

Part sermon, part anatomy lecture, this passage sounds as if it were written expressly for Carter, who, of course, knew these places well. It also contains the key to understanding a puzzling diary entry, which, in turn, unlocks a secret to understanding the young H.V. The entry comes in the first pages of Carter's first diary. On Sunday, January 7, 1849, the seventeen-year-old reports going to chapel twice and spending time reading Paley's *Natural Theology*. On the left half of the page, he jots, "Acquired a <u>new idea</u> from it," underscoring his excitement. And just to the right, he places a quote I recognize from Ecclesiastes, the most dour book of the Old Testament: " 'With much <u>wisdom</u> is much grief and he that increaseth his knowledge, increaseth troubles.' " The one thought is shoved to the left side, the other to the right—*For* and *Against,* I realized. But why? Why the opposing thoughts?

Like Paley, he could see the divine in the body. But here's the rub, as Carter might say. Even at this early date, he knew too much—too much about human anatomy—not to also see the imperfections, the *flaws* in the body's design (for instance, the single passageway that both food and air share, which can result in choking). Hence, for H. V. Carter, the conflict between faith and knowledge was embodied in the body itself. The Christian within him could see what Paley could see. But the anatomist knew better.

Seven

AFTER THREE WEEKS OF CLASS, I AM SERVING AS A DEMONSTRATOR of anatomy.

My duties are quite different from Henry Gray's when he was demonstrator of anatomy (one of his many roles at St. George's). He would stand in front of the class, behind a cadaver, showing students the parts of the body being described by the lecturer. The anatomy I am demonstrating, by contrast, is my own.

With my shirt off, I have assumed what is called the anatomical position: standing upright with feet together, hands at the side, palms forward, with eyes directed toward the horizon, or, in this case, toward the Golden Gate Bridge, visible due north out the lab windows. The anatomical position, used in all medical disciplines, is the standard body position for referring to the location of any structure. In this pose, the subject becomes an object.

Following Dr. Topp's directive, the eight PT students grouped around me have to find the "bony landmarks" of the surface anatomy of the upper body, all fifty of them. Some are clearly visible, such as the shaft of the clavicle or the spine of the scapula (the collarbone and the sharp edge of the shoulder blade, respectively), but most need to be palpated. The gloves come off, the better to feel what is just under my skin. An easy one to find right away, on me as well as on most people, is the C7 spinous process. Translation: the bump at the seventh cervical vertebra. To most of us, this is the knobbiest knob you feel when massaging the back of the neck. To a physical therapist, it is the starting point for locating any of a patient's thirty-two other vertebrae. Counting upward, there's C6 to C1, while downward from C7, the remainder are grouped by region, with twelve thoracic, five lumbar, five sacral, and, finally, at the tail-bone, the three to four coccygeal vertebrae.

Just as landmarks are handy when one is driving—*Take the second right after the movie theater,* for instance, or, *If you pass Millie's Diner, you've gone too far*—the landscape of the human body is more readily navigable for its markers. One of the most clinically important is the sternal angle, a slight ridge of bone near the top center of the ster-num, or breastbone. If you've ever pointed at yourself while asking someone, "Are you talking to me?" you have probably pointed right at it. Its exact spot is most easily found by first slipping your finger into the groove at the very bottom of the throat, which is the very top of the sternum. This is the jugular notch, and the tip of the fin-ger seems to fit perfectly in it. Now slide your finger down the ster-num about 1½ to 2 inches (4 to 5 centimeters) to the first tiny bump

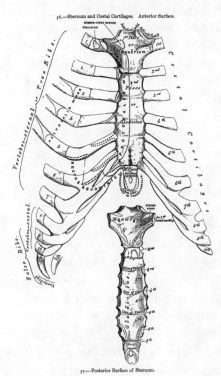

56.—Sternum and Costal Cartilages. Anterior Surface.

57.—Posterior Surface of Sternum.

or ridge. You have found the sternal angle. Just behind it, along the same horizontal plane, the arch of the aorta begins its arching, the tra-chea splits into the bronchi, and, deeper still, the thoracic vertebrae 4 and 5 are situ-ated.

A landmark to other clini-cal landmarks, the sternal angle is indispensable for correctly identifying the ribs, which are counted top down, with twelve on each side. The first rib, embedded under the clavicle, cannot be palpated. The second, how-ever, lies to the right and left of the sternal angle, though the bump can be subtle.

From here, moving downward and sideways, you can count ribs 3 through 10, all of which connect to the sternum via a small portion of cartilage. The last two ribs on each side are unique in that they end roughly halfway around the torso. These are the "floating ribs," and their tips can be felt just above the waistline.

Now, repeat eight times, without squirming, and you'll have an idea of what my morning has been like. As it turns out, my rib cage is good for counting because I don't have a lot of body fat. My arteries are easy to palpate, veins easy to see. The students tell me I have a very well defined bicipital groove and impressive biceps, too. But compliments gradually give way to a sharper truth: my body serves as a primer on being middle-aged.

Both my scapulas wing a bit too prominently, a possible sign of weakening serratus anterior muscles. My collarbones are uneven, probably from shouldering my gym bag on my left side for the past twenty-odd years. And my right shoulder joint makes a clicking sound as it rotates, apparently an early sign of arthritis. Once this is discovered, the entire class descends and I become the morning's star attraction: *Come, see the man who clicks!* I am glad we are not going to move on to the lower limbs today—the hip joints, knees, and arches—where I have more serious problems, thanks to years of running and working out. Instead, we shift from the living body back to the dead for the remainder of lab.

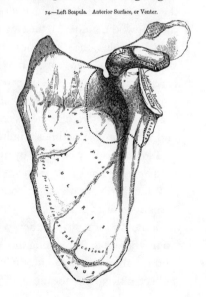

74.—Left Scapula. Anterior Surface, or Venter.

We have to perform a blunt dissection, meaning that the focus is on the destination, not the getting-there. All that's required is a large scalpel blade, a deep breath, and an extra-deep cut. It takes half a minute, maybe less. For this particular dissection, you just slice straight across the crease at the crook of the arm (*you* being *me*, in this case, with Kristen, Sam, Kelly, and Cheyenne looking on), then make small perpendicular incisions at either end, peel back the flesh, and—

"*Wow.*"

The five of us are a choral group of exclamation, though we know only one song. We sing it again: *Wow.*

Peeking through the crevice of mud-colored flesh is a glint of ivory, our truck driver's elbow joint, our destination. I finger away the fascia till the joint capsule is fully exposed, then snip away its papery protective membrane.

Even in a long-dead body, a joint remains a thing of beauty. It is smooth and shiny and still gleams with synovial fluid, the clear substance that keeps the joint lubricated or, if you will, greased. In *Gray's Anatomy*, Henry Gray describes synovial fluid as "glairy, like the white of an egg," and "having a slightly saline taste," a phrase that begs a question that is perhaps best left unanswered: did he actually put a dab on his tongue?

I take a step back as my lab mates take turns digging a little farther, exposing the three long bones that meet, or "articulate," here at the elbow: the humerus, the single bone of the upper arm; and the radius and ulna, the two bones of the forearm. Like all joints in the skeletal system, the elbow joint is not a part per se but a *place* where parts come together. However, unlike the simple hinge of a knuckle joint, for example, the elbow joint is tripartite, a three-way intersection of bone. Here, the radius and ulna articulate with each other and, separately, sometimes simultaneously, with the humerus.

A joint is usually powered by muscle and nerves. This one is powered by Kelly. She pronates the cadaver's arm, and we are all floored.

Pronating is rotating the forearm so that a palm facing upward turns downward. A person does this countless times in a day without a moment's thought, as when you turn your hand over to look at your wristwatch. Yet, seeing the inner workings of this simple act is nothing short of profound. Kelly turns the arm again. In a single, perfect movement, the head of the radius spins in place on the humerus while the shaft rolls over the ulna. They return just as gracefully to their side-by-side position during supination, the reverse action, which again turns the palm upward. That a bit of life seems to linger in this dead body is surprisingly unghoulish.

We all take a turn, rotating the arm as well as flexing it in a modified biceps curl. Just making the movement occur, though, is not enough. Each of us is compelled to place a gloved finger directly in the elbow joint as the bones twirl and glide, to feel what can only be felt from the inside: movement at its source.

Dr. Topp, overhearing our excitement, comes to take a look. "Even after all these years," she leans in to me and says, "I still find that really cool, too."

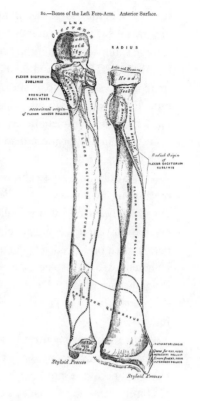

80.—Bones of the Left Fore-Arm. Anterior Surface.

I look around and see jaws dropping at every table. Celeste and her group have exposed the shoulder joint, a ball-and-socket that has pulled them into its rotation. One table down, it is circumduction of the wrist. Two over, it is pronation of the opposite forearm, the elbow dance in reverse. Students stroll around the room, table to table, watching, making, and feeling the movements. Everyone here, the living and the dead alike, has become a demonstrator of anatomy.

IF NOT IN the classroom demonstrating, he might be found in the lab dissecting. And if not in the lab, then at the hospital, whether in surgery or in the morgue or in the Dead House, or maybe in a meet-ing, being by now a member of the board of governors of St. George's, not to mention the London Pathological Society and the Royal Medical Society and the Hunterian Society. And with the few spare minutes left in a day, in early October 1852, Henry Gray could most likely be found in his office at home. But do not disturb him. He is trying to complete his treatise on the spleen, a project he has been working on for more than two years now, and it is due in less than two weeks.

This would be Gray's submission for the Astley Cooper Prize, a prestigious award based on a dead man's unusual posthumous re-quest. As mandated in the will of Sir Astley Cooper (1768–1841), every third year a judging panel would accept manuscripts of origi-nal research on a predetermined anatomical part, pulled from a list he had drafted. The spleen was the current topic of inquiry. The not insignificant cash award to this triennial prize, paid from a sizable en-dowment, was £300. The winner would be announced the following July.

Cooper, a highly regarded English anatomist, surgeon, and profes-sor, had died when Henry Gray and Henry Vandyke Carter were just boys. But he was responsible, in a sense, for bringing them together. Had it not been for Sir Astley's prize, after all, Gray might not have found himself in need of an artist, back in June 1850; Carter might not have found himself offering his services; and, ultimately, the two Henrys might never have become friends.

There is no better gauge of a friendship, I believe, than the ability to do nothing together, and, as Carter's diary testifies, the two had little trouble in that department. They would often while away quiet afternoons together, and one day Gray even coaxed Carter into doing something completely out of character: the two men stepped out of the hospital doors and just kept walking. Once they got to Chelsea, they took a leisurely boat ride on the Thames, eating up a whole half day. Carter, in other words, had played hooky! He nor-mally fought against "idleness" and berated himself for each mis-spent minute, yet he never portrayed being idle with Gray as wasted time. Gray felt likewise, no doubt; in H. V. Carter, he had met a kindred spirit.

Despite the differences in age, religion, and family background, these two shared a keen interest in medicine and science, of course, but also a passion, verging on the macabre, for dissecting. For instance, even after Gray completed his one-year stint as postmortem examiner (1848–49), he continued on in an unofficial capacity. Quite often, Carter attended these autopsies as an observer, and at least once that I can verify, he assisted Gray in a PM exam. For both men, a postmortem was a dissection with a puzzle built in: *What went wrong here?* Sometimes, in addition, the examination yielded an unexpected anatomical treasure, such as a heart with four rather than three cusps to the aortic valve, an astonishing anomaly.

As for Carter, his dissecting proclivity was never more pronounced than during summer vacation back home. On top of fishing with his brother or walking and people-watching with his sister—seaside Scarborough swelled with vacationers ("trippers") this time of year—he would spend many an hour performing dissections. Not on cadavers, mind you, but on local creatures, frogs and fish and such. In one nigh-maniacal marathon, he first dissected a single snail, then five more, then collected an additional half dozen for future disassembly. Oh, and p.s., "Snails not easily killed!" he noted boyishly in his diary. Though his relentless dissecting seems almost comical (and I can only imagine what his parents thought of this pastime), it also shows Carter's seriousness of purpose, as he was, in effect, teaching himself comparative anatomy, the study of the similarities and differences in the structure of living things.

Following the 1850 summer break, he returned to London and performed what would become a ritual over the next few years: he immediately checked in with Henry Gray. I get the sense that seeing his friend after time away was a welcome, and perhaps even necessary, way to transition back into frenetic London life. Gray, like an affable older brother, clearly had a steadying effect on the anxious, temperamental Carter, and he was always encouraging and respectful of his talents.

Carter's involvement with the spleen project came in two bursts. He created twenty-three paintings and drawings initially, but it wasn't until April 1852 that the senior Henry again needed his artistic skills. This time, the focus was the spleen in animals. These drawings were done primarily at the Royal College of Surgeons, a two-mile (three-and-a-quarter-kilometer) walk from St. George's, where Carter worked from the school's extensive collection of preserved animals. And, no doubt, his knowledge of comparative anatomy proved useful, especially since Gray, by his own admission, had little experience in this branch of anatomy. But what made these drawings unlike any he had ever produced was that, once the batch was completed near the end of June, he got paid. Better than finding a fourth cusp on an aortic valve, this was H. V. Carter's "<u>first</u> professional engagement" as a medical artist, and Henry Gray had made it possible.

Actually, make that "Henry Gray, F.R.S."

Gray had been bestowed those three little letters just a few weeks earlier. On June 3, 1852, he was elected a Fellow of the Royal Society, a singular honor for a man of just twenty-five. His candidacy had been supported by a long list of Fellows, yet it was really Gray's own work that had provided his highest recommendation. On earlier occasions, two of his scientific papers—one detailing original research on the development of the human eye, another on the spleen—had been read before the assembled Fellows, then discussed by the group, an experience that must have been as heady the second time as the first. Both papers were accepted for publication in the society's prestigious journal, *Philosophical Transactions*. And shortly after being named a Fellow, Gray received a £100 grant from the Royal Society, to be used toward completing his investigations into the spleen, including, presumably, paying for his artist's efforts.

Carter had recently acquired an impressive set of initials as well. On May 21, the day before his twenty-first birthday, he became a Member of the Royal College of Surgeons, having passed the extensive entrance exams. Now certified to practice surgery, Henry Vandyke Carter, M.R.C.S., still had to obtain an apothecary license in order to become a full-fledged physician. The apothecary exams would be coming up in October.

Carter, who had completed his required apprenticeship with Dr. Sawyer earlier in the year, had also moved out of the Sawyer home and acquired a new address. Joined by Joe Carter, who had come to London to study art and, if H.V.'s rantings are to be believed, to torment his older brother, had moved into an apartment on Upper Ebury Street. Though he now lived much closer to Henry Gray's home, gone were the idle days of old. The demands on their time had become too numerous, Carter with his studies and exam prep, Gray with his many professional obligations and the looming deadline for the Astley Cooper Prize.

When that day finally came, it received a quiet mention in Carter's diary. "See Gray," he notes on October 13, 1852. "He has just finished his subject." If Carter sounds weary, it is for just cause. Six days earlier, he'd "passed the <u>Hall</u>," meaning he had earned his apothecary license; all his formal schooling was done. And six days hence, he would be setting his London life aside and leaving the country.

OCT. 1852. "19 Tu. Last day in town. Today got passport & ticket. Back just in time to get things together in 2 carpet bags—one Mrs. Loy's—leaving many important things, amongst others, my Bible, which much regret—"

Mrs. Loy? That's his landlady.

Why the hasty departure? And where did he go? Well, again, Carter himself does not explain. As is the diarist's prerogative, he doesn't have to. The same rule does not apply to the letter writer, thankfully. Letters demand a narrative and very often provide an explanation. I therefore returned to the Carter archive, placing an order for 3 of the 116 surviving letters from H.V. to his sister Lily, who was nineteen months his junior. These three bear the postmark "Paris." I requested scans this time, this time being impatient, and within days, via online delivery, I, like Lily Carter almost 153 years ago to the day, found myself opening mail from H.V.

October 23, 1852
Hotel de Seine
Rue de Seine
Paris

My dear Sister,
    This is the first quiet evening I've had since I left town and I take occasion of it (as the French might say, if they spoke English) to quiet your apprehensions at home, to satisfy, in a measure, your curiosity.

But before he continues further, a caveat:

    Do not anticipate, dear Lily, a detailed account of all I've seen and heard, nor yet, a chapter of horrors and oddities: what I write is meant for a succinct narration of facts and observations—so now to begin.

Fortunately, Carter does not stick to this game plan and proceeds with lively detail about his voyage by steamer and rail to Paris. On his first night there, two fellows he had met on the journey took him out on the town—and oh, Lily, what a time! "We supped at a grand café, a la Française, everything in great style such as you have never seen." Then, in a sentence that does not come up for breath, he tells how the three strolled in the Palais Royal, "a most extensive pile of splendid buildings with a square and fountain and gardens in the middle, and gorgeous shops around, colonnades, and arcades all lit up in the most brilliant manner and crowded with chattering gay French folk—the whole is a *tout ensemble,* certainly not equaled in London, or the world." Breath. "We were charmed."

He seems almost drunk on the details, and it is somewhere between his descriptions of the tree-lined boulevards with their magnificent houses and his sharing his plans to go to the Louvre the next day that I recall why I went to Paris for the first time at his same age: to *see* Paris. That is reason enough, if not reason alone, to pack one's bags. But Carter, it turns out, had also come to Paris with letters of introduction and a larger purpose: the man who had just finished his studies was actually continuing them.

As becomes clear in his second letter, he arrived in Paris just in time to get settled and make the opening ceremony for the winter session at the renowned La Charité Hospital medical school. Already, Carter had resumed a familiar routine. "In the morning I go very regularly to one of the great hospitals where the physicians and surgeons usually begin to visit at 8:00 or soon after. Then comes a lecture—a 'clinique,' we call it—then a walk to the 'Laiterie' (breakfast), where we arrive with a good appetite. In the middle of the day, I'm engaged at the lecture and dissecting rooms." In addition, he regularly goes on rounds and attends lectures at Paris's famed Hôpital des Enfants.

Part of me wants to say that making this trip was a very smart move on Carter's part, a way to gild his résumé (not to mention, improve his French). But I know better. I have glanced ahead. And his daily entries soon leave no doubt that something else is going on here, something he would never share with his sister.

He had forgotten his Bible but brought all his demons with him. Like a storm that's suddenly changed direction, Carter's crisis of the soul has shifted from doubts about faith—the impetus for *Reflections,* a volume that is tellingly silent during this Paris period—to overwhelming anxiety about his professional prospects. Now that he has obtained his diplomas, he must make the transition from student to practitioner, yet Carter sees nothing but difficulties ahead. Rather than face them, he is in Paris, a fugitive from his own future. As he writes on New Year's Day 1853, "The tolerable success and éclat of student's progress at St. George's is over. Then, knowledge was my sole aim; now, I must think of a livelihood."

His is not an uncommon dilemma for a new graduate, but I suspect he is operating under the misconception that learning must end with earning a living. Though his "love of science and the higher branches of the profession" remains steadfast, Carter confides that he has "no interest whatever in the Profession." He does not want to *be* a John Sawyer, a G.P. with an apothecary and an apprentice and so on. At the same time, Carter feels he still lacks the self-confidence, the inner oomph, "to strike high and risk the consequences," meaning, to be an innovator such as Henry Gray, for instance.

"Perhaps [I] ought then to be content with a lower station," he tells himself, "yet, and here seems the rub, my ambition is but just enough raised to cause inquiet. . . . This is the poison." And the poison paralyzes.

He has, however, set into play a possible escape plan. Just prior to leaving for Paris, he had made a number of discreet inquiries about becoming a surgeon aboard a "packet," or small cargo ship, traveling back and forth between England and India. The colony had become a major market for English goods, and hundreds of companies were operating vessels with fully staffed crews. In fact, one of the surgeons at St. George's had promised Carter help in getting a "surgency" with an outfit called the General Screw Steamship Company. Such a job must have sounded far more exciting to him than hanging out his shingle back in London or Scarborough.

That the idea of leading a more adventurous life would have appealed to the twenty-one-year-old is reinforced by a rather large clue he left behind. It comes on the very first page of his new daily diary, bought and begun in Paris. On an otherwise blank page, Carter affixed an elegant calling card bearing only a name:

J. BELLOT, LIEUTENANT DE VAISSEAU

It meant nothing to me until I returned to his first letter to Lily. Of course! Bellot was the name of one of the fellows who had taken Carter out on his first night in Paris, "a Naval English-looking young officer," he'd told his sister. In point of fact, Bellot was not English, as Carter soon discovered. But that was the least of it. After dinner and a good many glasses of wine, the men began discussing the latest news on Arctic exploration, and, as H.V. explained to Lily, "I burst into admiration of a French officer who accompanied the last expedition." This man's story had been in all the papers back home; he had become a hero to the people of England by volunteering in the search for Sir John Franklin, a celebrated English explorer and sea captain who had disappeared in the polar regions.

As Carter told his sister, he'd gushed and gushed about this brave French officer till, finally, "my friend stopped me and said he could not hear <u>himself</u> so praised—he was the very man," Joseph René Bellot! (the "Bellot of the Papers," as Carter would call him)—but so modest that no one would ever have guessed. "And we found him out by chance."

Next morning, Carter and Bellot breakfasted together, then spent the day walking around Paris—the Tuileries, Place de la Concorde, and other sights—till, finally, the two parted ways, Bellot having his own journey to complete. The young adventurer left behind an indelible memory and a calling card that Carter would later put to use when starting his new diary on New Year's Day. The name on the card would serve, in effect, as the epigraph for the next phase of his story—The Life of Henry Vandyke Carter, Volume 2—setting a bold tone for what was to come.

PART TWO THE ARTIST

Man is only man at the surface.
Remove the skin, dissect,
and immediately you come to "machinery."

—Paul Valéry (1871–1945)

Eight

SOMEDAY I MAY WONDER WHERE I FOUND THE NERVE TO DO this. It is just one hour after morning coffee, and I am helping to perform what is awfully close to a decapitation. We have turned our cadaver onto its stomach and have propped the chest on a block so that the head nods down, leaving the neck a clean downward slope. This is whiplash terrain, the thick, powerful muscles that help support the head, and I slice right through the three main ones: the longissimus capitis, the semispinalis capitis, and the splenius capitis (*capitis* meaning "head"). Kelly, the gang, and I take a few minutes to examine each fleshy cross section, then plunge ahead. Our ultimate goal is to dissect C1, the very top vertebra of the spinal column, a site of catastrophic neck injuries. Commonly called the Atlas, for the man of myth who held the world on his shoulders, this deeply embedded vertebra serves as the base for the globe that is the head. To reach it, we need to tunnel through several more layers of muscle.

I can see C4, like a subway token at the bottom of a meaty purse, but rather than stopping to finger it, I slice northward about six inches (fifteen centimeters). Now, Kristen, facing me on the opposite side of the cadaver, makes a horizontal cut across the base of the skull, connecting the backs of each ear. Where her line meets mine, we each begin peeling back the triangular flaps of scalp. Just like on my head, our truck driver's hair is buzz-cut (as are all the cadavers'), to make tasks such as this easier. The skin feels as tough and bristly as animal hide.

Adding to the discomfort of the day, the school's maintenance department is testing the ventilation system on this hot August morning, so all the windows are to remain closed for the duration of lab.

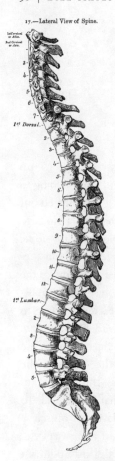

17.—Lateral View of Spine.

1st Cervical or Atlas.
2nd Cervical or Axis.
3
4
5
6
7
1st Dorsal.
2
3
4
5
6
7
8
9
10
11
12
1st Lumbar.
2
3
4
5

In the still, thick air, it seems as if *we* are being put to the test rather than the air-conditioning. Everyone is sweating, making it smell like the gym of the dead. Pulling at the pocket of my scrubs, I make a poor man's fan of my shirt-front. I am truly seeing the wisdom in the centuries-old policy of dissecting only during the winter months.

Think of this as an archeological dig, I tell myself, trying to remain positive. True, this is probably more unpleasant than sifting dirt under a noonday sun, but at least we're assured a discovery or two.

"There's C3," says Cheyenne over Kristen's shoulder.

"And here's C2."

We are just about to enter the suboccipital triangle, an area of dense muscle beneath the backmost lobe of the brain, when—

A throat clears.

Like a raven-haired cat, Dr. Topp has suddenly materialized at the foot of our table. "Today is your day," she says, and we know exactly what that means: a pop quiz, of sorts, with an anatomical twist. The group has just thirty minutes to put together a class presentation on the "functional anatomy" of a specific movement. As for the topic, that's up to Dr. Topp. Last week, she had Casey's group analyze "deep respiration with upper extremities fixed," meaning the classic just-finished-a-marathon position—bent over at the waist, hands propped against the knees. Two weeks earlier, Robyn and her team were given "the iron cross," that astonishing strength maneuver performed by male gymnasts on the stationary rings. Another day, it was the mechanics of a yawn.

And this time: "I want you to dissect . . ."—Dr. Topp pauses a moment—"a push-up."

Off come the rubber gloves as Kelly, Kristen, Cheyenne, and Sam head to the big chalkboard at the back of the room. As each student's grade hangs in the balance, I will serve only in a support capacity. I pull the cover over the cadaver before joining them.

152.—Muscles of the Chest and Front of the Arm. Superficial View.

Having spied on other groups as they prepped their presentations, I know that my table mates have a lot to do. They must figure out the exact sequence of muscles, nerves, and joints used in executing their assigned movement, which is anything but a simple task, especially given the half-hour time constraint. On the other hand, the time constraint is good training, forcing each of them to think on their feet, just as they would when assessing a new patient.

The four of them decide to break the assignment in two. Kelly and Kristen claim the upward motion of the push-up—the push away from the ground—while Cheyenne and Sam take the downward movement, which sounds like an excellent plan but rapidly proves otherwise. They realize that the two actions don't happen in isolation and each duo will be doing too much overlap. So, scratch that. The group regroups. Just as the body works together to create a movement, the team must work together to break it down.

They start from the top, literally. In the push-up start position, they determine, holding the wrists stable requires four carpal extensor and flexor muscles and three spinal nerves. Maintaining slight elbow flexion relies on the triceps brachii, anterior deltoid, serratus anterior, and several thoracic muscles, plus the radial, axillary, and long thoracic nerves, as well as spinal nerves C5 through C8 and T1. Holding the neck steady and level requires another ten nerves and muscles.

At eighteen minutes and counting, the chalkboard is a madman's cell wall.

As they focus on the next step, lowering the body toward the ground, they get stuck—*really* stuck—unable to agree on which back muscles are, and which are not, involved.

"Maybe I should do a push-up for you," I volunteer.

All four look at me as if I were a cooling breeze.

Ten push-ups later, things start clicking, and not just from my shoulder joints. The lats, the traps, the pecs, both major and minor, all come into play. Click, click, click. The scapula, humerus, and glenohumeral joint join the list—

"And we can't forget to mention gravity!" Sam interjects, more impassioned than I have ever seen him. To the scritching sound of chalk on a blackboard, the group has transformed from an anxious lot to a confident one. I have no doubt whatsoever that they will nail their presentation. And, sure enough, they do.

Though I hate to play favorites, I must concede that my favorite in-class presentation does not come until two weeks later. It is "the queen's wave," as analyzed by Adrienne and company. Watching these four energetic young women deconstruct this signature of royal reserve is a delight. Somehow, the white lab coats and ponytails add to the charm. I also find the movement itself fascinating; it barely squeezes into the dictionary definition of the word *wave,* for the fingers do not wag. Rather, the queen's wave is an upright hand performing what looks like a slow stirring of the air.

Such subtlety does not come simply, as the team explains. In one motion, the clavicle elevates, the scapula rotates, and the shoulder abducts, all in service to the arm as it rises fluidly into the air. At the same time, the hand cups and the forearm supinates ever so slightly. The queen, remember, waves with the back of her hand rather than the palm. Of course, the greeting is impossible without a great interplay of muscles and nerves, but what really makes this movement majestic, it strikes me, occurs in the carpal region. In other words, it's all in the wrist, which must be held perfectly still, as if it were an anatomical exemplar of monarchical stability. This is where a wave becomes a wave becoming of a queen.

JUST AS A chance encounter with Joseph Bellot "gave an éclat" to H. V. Carter's "entrée into Paris," as he had told Lily, a brush with royalty brought his stay to a memorable close. On Sunday, January 30, 1853, two days before Carter packed his bags for home, he joined the crowds lining the streets for the grand procession of Napoleon III and his betrothed on their way to Notre Dame to be married. The forty-four-year-old emperor, a nephew of Napoleon Bonaparte, had chosen as his bride a Spanish-born beauty named Eugénie, who was almost twenty years younger.

"Whole scene exciting. All Paris out," Carter reports in his diary.

The festivities continued into the night. The City of Light shone as never before, with the Place de la Concorde aglow with "electric lights" and "illuminations very brilliant everywhere." Even so, it was Carter's glimpse of Empress Eugénie earlier in the day that still burned most brightly: "Nose aquiline, chin small, upper lip [a] little curved," he notes, with an artist's eye for detail. Watching her greet the throngs, Carter had not seen joy in the twenty-six-year-old's lovely face. Instead, "expression quiet and resigned."

Drawing upon this memory, he would create a portrait of Eugénie shortly after getting home. But first things first. As per tradition, he marked his return to London life by checking in with Henry Gray. They met up in the Dead House and swapped stories, Carter reports in his diary, he of his Parisian adventures and Gray of his own encounter with royalty of sorts. No less a luminary than Dr. Caesar Hawkins had paid him a recent visit, Gray told his friend. Hawkins, who was known around St. George's as "the Emperor," not for lording about the place but for the great respect in which he was held by hospital staff (not to mention his imperial-sounding first name), had heaped upon Henry some glowing praise: your anatomical preparations, he told Gray, "are a credit to England." (Hawkins was certainly qualified to make such a pronouncement. Not just one of the hospital's chief surgeons, he was also the newly retired president of the Royal College of Surgeons and would one day be named sergeant-surgeon to Queen Victoria herself.) Though the doctor's compliment had come a couple of days earlier, it is clear through Carter's recounting that his friend was still floating on air.

Gray had also just been named head of the Anatomy Museum, a well-deserved promotion in Carter's eyes. Gray "is well-placed as curator and fully alive to his advantages," he writes that evening. "Envy him." Not that he was wallowing in self-pity, mind you. On the contrary, Carter, who'd scarcely had time to resettle into his old flat, was already taking decisive steps in shaping his own destiny. While still waiting to hear back from the steamship company, he had decided to pour his energy into his artistic work. His first move was to assemble a portfolio of his anatomical drawings and paintings—a "specimen" of his work, as he called it. As a student, he'd never had the time or inclination to promote himself as an artist for hire, or the financial need. Sure, occasional jobs had come to him through Gray and other faculty, but, simply happy to contribute, he had nearly always done the work for free. As of now, that policy would have to change.

Carter, ever the anxious soul, had not made this decision lightly. For guidance, he'd turned to Prescott Hewett, one of a handful of father figures in his life. It seems that the question he brought to Dr. Hewett was not whether he *could* make money as an anatomical illustrator but whether he *should*.

Propriety told him no. Wouldn't he be "encroaching" on other artists' territory? This concern stemmed from H. V. Carter's upbringing, I believe. As the child of a working artist only now finding fame, he knew firsthand what a struggle it could be to make a name for yourself, to become established. He would not want to threaten another artist's livelihood or, for that matter, to be viewed as a dilettante—a physician who simply dabbled.

Dr. Hewett absolved Carter of these concerns, assuring him that drawing was in fact a "perfectly legitimate" enterprise. *By all means, young man, use your talent!* And like a racehorse on Derby Day, he was off and running.

True, he did stumble right out of the gate—Carter's first job prospect fell through, leaving him "disappointed and put out"—but he recovered quickly. The next day, in fact, less than two weeks after returning from Paris, he showed his portfolio to three separate doctors, all of whom promised him work.

"Hence," he writes with brio that night, "have regularly set up as a <u>Medical Artist</u> and have little doubt, D.V., (*Deo volente,* or, God willing) [that] in a little time might make it pay well." Carter sounds full of confidence, as well he should be, and yet he cautions himself never to lose sight of his top priority: "The exercise of the profession is the chief end . . . of [my] medical education," and "the artist's position is but subsidiary." He adds, *"Pro tempore!"* meaning "for the time being."

After four weeks back in town, he earned his first fee, £4.7s for several days' drawing for a Dr. Heale, an encouraging amount. Heale offered additional work, as did Gray, but already Carter was growing restless. "[I] am constantly feeling want of fixed and full employment." He was not willing, however, to take just any job.

Twice, he is offered a full-time "assistancy" position, the kind of entry-level job most freshly minted doctors had to take, and he turns them down. One offer had actually been forwarded by his father from a medical practice in Scarborough, no doubt along with a fatherly nudge. "Hardly know how to treat offer," he writes, showing a moment's hesitation before again standing firm: "I am fit for a higher office." Of course, I know that he knows that he still has a dream job in mind. Carter remains hopeful for a berth on a steamship.

One day in early March, he stops by the London office of the General Screw Steamship Company. Though his name has apparently been "on the list" for an interview since the previous October, "appointments are slow," he is told. But this does not sit well with an eager young man. After weeks of forced patience, Carter writes a letter to the company's chief officer, asserting his earnest "desire to enter service." He posts the letter on the nineteenth of March, the very same day, coincidentally, that an ad he had purchased appears in the distinguished medical journal *The Lancet.*

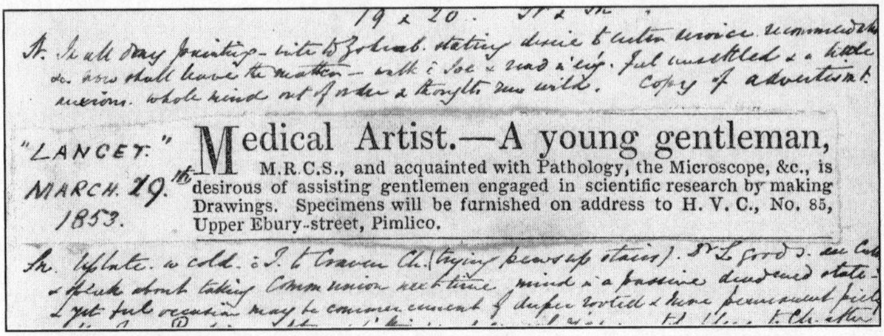

Medical Artist.—A young gentleman,
M.R.C.S., and acquainted with Pathology, the Microscope, &c., is desirous of assisting gentlemen engaged in scientific research by making Drawings. Specimens will be furnished on address to H.V.C., No. 85, Upper Ebury-street, Pimlico.

This was his calling card, presented on the newsprint equivalent of a silver platter, to the entire medical community of London. He had fussed over each word, checked the proof for typos, and when the ad came out, pasted a neatly clipped copy into that day's diary entry. For him, the nineteenth was a day to imagine, and savor, all the fantastic possibilities ahead—the flood of responses to his ad, the encouraging reply from the steamship company, the subsequent interview and job offer, his first journey out to India, and on and on. The future looked bright. With the turn of a page, however, come the gray clouds. The steamship company informs Carter that his "age and inexperience" make him an unsuitable candidate for a ship's surgency. Adding to his disappointment, it soon becomes clear that *The Lancet* ad fails to generate any new work. But why? Was the advert too genteel?

Carter doesn't give up but rather changes focus. He now sets his sights squarely on a Studentship in Human and Comparative Anatomy offered by the Royal College of Surgeons. This was essentially a full-time internship of two years' duration and was awarded every June to the winner of a highly competitive qualifying exam. This meant he had just over two months to prepare.

Already licensed to practice surgery, he was not lacking in impressive academic honors and credentials. In fact, he had returned from Paris with six new "certificates" attesting to the specialized work he had done in hospitals there. And the studentship would not give him the kind of on-the-job experience the General Screw Company thought he was lacking. One could even argue that the position would be a step backward for him. Indeed, while casting this last line of thought, I believe I found the likeliest explanation. The previous June, just a few weeks after earning his M.R.C.S., Carter had come in second in the studentship exam—so close!—but second place got you nothing, not even a certificate. This time around, he would redeem himself and claim the crown.

And June 14 was the day. "I was called in, as [the] successful candidate for [the] Studentship of Anatomy to [the] College!" he tells his diary, skipping his articles in his excitement.

One of the very next to hear the news was Henry Gray, whom Carter visited at work. Not only pleased for his friend, Gray offered further encouragement, suggesting that he now go for the Royal College of Surgeons' Triennial Prize. Gray himself had won this award four years earlier for his study of the nerves of the human eye. *You'll have access to the lab and the library, so why not do some independent research?*

"Might—shall see," Carter notes, and his reluctance is understandable. This would be a major undertaking, requiring an effort comparable to a master's-level dissertation, and he has already got a lot on his plate with the upcoming M.B. (bachelor of medicine) exam. (He must pass the M.B. before then working toward his final accreditation, the M.D.) Carter being Carter, I am sure he was just worrying about surviving his first day on the job, whereas Gray, also true to form, was thinking years ahead.

They pick up the same conversation several days after Carter has begun the studentship, and this time, the subtext is much clearer. Perhaps Gray knows that the position won't be as challenging as Carter might want and, knowing his friend as well as he does, believes he needs a fixed goal to work toward. Gray presents his concern in "a considerate and encouraging way," but Carter is still in wait-and-see mode.

Already, though, Gray seems to be right. Over the previous six days, Carter had done nothing beyond a little dissecting and had yet to even see Mr. Queckett, the professor he was meant to assist. One person he had not been able to avoid, however, was an insufferable chap named Sylvester, the first-place winner in last year's exam. Now the senior student to Carter's junior, Sylvester sounds like the anti-Gray. "Hasty and spiteful," he had made nasty little digs at Carter, telling him, for instance, that the dissections he had done for the exam were inferior to another candidate's.

Making an uncomfortable situation worse, Carter is under the impression that Caesar Hawkins and the other higher-ups at St. George's are upset with him for taking a studentship at what is essentially a competing school. But this is sheer poppycock, as Lily might say, another instance of Carter's letting a neurotic sense of propriety get the better of him. Henry Gray, to whom he turns for counsel, assures him likewise.

Always there to buck him up, Gray is the ever-friendly port in a storm, whether said storm is imagined or not. He makes everything seem, well, not easy by any means, but *within reach,* so long as one works hard. To someone as impressionable as Carter, however, there is a clear distinction between an everyday role model and a paragon, a being kissed by destiny, and, as of July 25, 1853, Henry Gray appears to have crossed over. Of his friend, Carter writes that evening, "Gray got [the] Astley Cooper prize—beating good men." He seems a bit incredulous, as if thinking, *How* does *Gray do it?* "Clever fellow," he cannot help but marvel.

Along with the £300 cash prize, Gray would be accorded an even greater reward. His treatise on the spleen had attracted the attention of a London publisher who would release it as a book the following year. In a small way, Carter shares in this victory because, of course, he had created the illustrations for the project.

Unfortunately, he doesn't hear the good news from Gray himself, for Gray has been "down in [the] country—[He] has been very ill." Ill with what, and how serious the affliction, Carter does not say, but, as I've learned through St. George's administrative records, Gray's illness was serious enough that he requested a leave of absence from his curator duties. Further, he arranged to defer receipt of the Astley Cooper Prize until he had recuperated.

Hardly two weeks pass before Carter loses another anchor in his life, his brother and roommate Joe, who suddenly falls ill and is bundled off to recuperate in Hull, with the boys' aunt, uncle, and grandfather. "Somewhat alarmed—choleric symptoms," he jots that night in his diary.

Joe, now eighteen, had rejoined H.V. upon his return from Paris to continue studying art, and the two shared expenses. "Joe and I get on pretty well," H.V. told Lily in a letter from February 1853, but then backpedaled. "His malpleasantries, however, stick to him like pitch and make me too sharp perhaps."

Perhaps? Lily surely got a giggle out of that. H.V. and Joe, polar opposites, annoyed the hell out of each other in a way only brothers who love each other can. Joe was free-spirited and fun-loving and did not seem to have a religious bone in his body. Mostly, what was on his mind was girls, at least from his older brother's perspective. As H.V. once commented to Lily, half in jest, "Joey's love for the young ladies is still as strong as ever, but I should like to know how he intends to support his wives."

While finding Joe "certainly very exasperating occasionally," H.V. also saw great artistic potential. The two often visited galleries and museums together, sharing in the splendid works in the royal collections—paintings by Titian, Turner, and H.V.'s namesake, Van Dyck, among others. What's more, just as their father had a penchant for giving impromptu art lessons, H.V. had taken it upon himself to give his brother "anatomical lessons," believing that a solid grasp of human anatomy was essential to Joe's education. "He won't make a good anatomist," Carter reported dryly to Lily after one such lesson, but that their little brother would one day be known as a great artist, he seemed to have little doubt.

And now that he was gone, H.V. missed him terribly.

"Feel very dull without Joe," he admits to himself—and to an empty apartment—after ten days on his own.

By the end of September, Henry Gray was back in town—and back to his old self. Within days, he had offered Carter a new challenge, asking him to create two drawings of a size and complexity he had never attempted. These huge drawings of the chest would be used in the classroom at St. George's, where Gray was now in his second session as lecturer of Practical Anatomy.

Carter agreed to the assignment, though not without private reservations. Rather than feeling idle, he now had the opposite problem: "Am too copious in plans," as he puts it, and that's putting it mildly. On top of rebottling hundreds of specimens at the college and helping Mr. Queckett prepare his histology lectures, he must squeeze in time to study for the M.B.'s. But at least he's busy. He really is happiest when he's busy—or, to be more precise, when he doesn't have time to be *un*happy. Before he has even finished the drawings, though, Carter gets slammed with a fresh set of demands: His uncle pays an unexpected visit to London, expecting H.V. to play dutiful nephew and host. Sylvester takes an indefinite leave from his job, leaving Carter to do double duty. And then another professor at the college recruits Carter to perform dissections for him, the first assignment being a whopper of a walrus. He is just barely keeping his head above water when, literally from the top of the world, shattering news arrives concerning Joseph Bellot.

"It is my melancholy duty to inform you . . . [that] he has lost his life," the story in *The Times* of London begins.

Although it reads like a personal letter of condolence, the Tuesday, October 11, story is actually a transcript of a dispatch filed from Her Majesty's Ship *North Star* at Beechey Island in the Northwest Passage. The twenty-seven-year-old French lieutenant had returned to the Arctic as part of a new English expedition searching for Sir John Franklin,[1] and things had gone terribly wrong, as the com-

---

1. The bodies of Franklin and his crew were finally found and exhumed in 1984, perfectly mummified in the Arctic ice. Forensic tests proved that the men had died not from exposure but from lead poisoning, transmitted via lead in the metal used for the canned goods they'd brought for the long journey.

manding officer reports. While making a perilous ice crossing with two shipmates, Lieutenant Bellot plunged into the dark waters of the Wellington Channel and drowned. His body was not recovered.

Though the dispatch had only just reached London, the tragedy had actually occurred eight weeks earlier. But as Carter and much of the city read the account, Bellot perished right there on the page.

Lieutenant Bellot was mourned in England and France and served in a small way to unite these historically antagonistic countries. Emperor Napoleon III took the unusual step of granting a pension to Bellot's family, while in England funds were raised to erect an obelisk near the river Thames to commemorate the Frenchman's role in the search for Franklin. *The Times,* following up on the story, ran a series of testimonials to the intrepid polar explorer. H. V. Carter submitted one himself, fondly recollecting his "accidental encounter" with Bellot in Paris. To gauge the full impact Bellot's death had on Carter, however, one needs to go back to his diary—actually, diaries plural.

Carter reports the news of "poor Bellot's decease" on the same date in both his daily diary and *Reflections,* using almost identical phrasing. This rare redundancy was no mistake, I believe, for Bellot's death was a blow to both body and soul, and Carter felt compelled to address both aspects of himself. In his daily diary, Carter then adds what reads like a non sequitur: "Am 5 ft. 11½-tall, weigh 10¼ stone." My first response to this was, *Thank you very much, H.V.* This was the first physical description he had provided, and as I had come across no photos or portraits of Carter as a young man, I was now able to begin picturing him. What a string bean! Though 4 inches taller than me, he was 10 pounds lighter (10¼ stone converts to 143 pounds, or 65 kilograms). And he towered over Henry Gray, who was five-two, tops. How odd those two must have looked if standing side by side— the one Henry tall and thin, the other stocky, dark, and gnomish.

By listing his measurements, Carter obviously was not trying to do historians a favor. He was taking stock of his corporeal self, I have come to see. The stark reality of Bellot's death struck deepest as Carter put pen to paper that evening, and he reacted much as one might after a bad traffic accident, say, when you step out of your vehicle and give yourself that kind of mental pat-down, making sure you're still all there. In the upswell of emotion, I'm not even sure he was fully aware of what he was recording on the page. After his height and weight he also notes his lung capacity (a "sound" 240 centimeters per inhalation), which suggests that he may have even gone in for a physical exam that day. In any event, I find this last detail almost poetic in its inference, as the drawing in of breath is the drawing in of life.

That the specter of death spooked him so surprised me. As an anatomy student, Carter had handled dead bodies for years, routinely carrying the scent of death in his hair and hands and clothes. As a surgeon in training, he'd certainly seen many tragic deaths, and he lived at a time when death by infectious disease, inoperable illness, or irreparable injury was common. But I suppose he also possessed the blitheness of youth, that sense of invulnerability that's not lost until you cross irreversibly into adulthood. With Bellot's death, Carter must have felt—and *recognized*—that cold breath on the back of his own neck, and now he couldn't shake the feeling. One night in his diary, for instance, he sounds panic-stricken when reporting that an acquaintance is "fearfully ill with typhus fever" and that another is already "dead of the same! Fearful—fearful warnings! Why am I left so intact?"

In this rattled state, he experiences a series of stumbles, some little, some big, but all hugely amplified in his mind. For instance, Carter misses an appointment with Gray one day, "from not being quite punctual, a failing of late with me." But a far more substantial failure soon follows: in mid-November, he bombs the M.B. exam— bombs it big-time. "Truly a week to be remembered in the history of my Studies," he writes, after the fact, "for have failed when about to grasp the last and highest scholastic distinction have aimed at." Unlike other friends and colleagues, Henry Gray shows great kindness in not even bringing up the subject of this "late failure," Carter reports.

How swiftly his circumstances had changed. I could still envision his face lit up in the grand illuminations of the City of Light. In less than the span of a season, though, Carter had become a different person, more mature but also touched with melancholy, as evidenced in one of his final entries of the year. The date is Tuesday, December 13, 1853, and he sits alone in the dissecting room at the Royal College of Surgeons:

"Through the kindness of Mr. Queckett, [this] room—otherwise so bare, unnecessarily so, of almost all necessary accommodations—is now furnished with a lamp to eke out these dark afternoons." In what seems like a fitting metaphor for his own changed perspective, the lamplight exposes aspects of the room he had never noticed before. "The mantel-piece looks desperately untidy," for instance, while "the massive beams overhead remind us of proximity to the skies (in some respects no bad position). We are at least, doubtless from best intention, placed in Attica—the abode of poets."

Alas, this illusion is dispelled, Carter writes, "when we approximate ourselves to the tank for injecting [filled with preserving fluid], by the unpoetical odours there abiding."

Fixing his gaze on the specimens of distorted fetuses and animals lining the shelves, he now sees something almost sinister within: "Look at the gaunt bottles, filled with huge deformities, some in mock derision of our best and earliest state—childhood. Infants without parts—and to cap all—the brat of a Chimpanzee so like any human offspring that one doubts one's nature," a reference no doubt to the emerging theory of evolution.

"All nature swarms around us," Carter adds sardonically, "the student is a very Adam! He is the keeper of a menagerie—of bones."

Nine

I F YOU THINK OF BONE AS BONELIKE—ROCK-HARD, INERT, THE prototypical caveman's club—you are mistaken. In the living body, bone is actually a dynamic tissue shot through with nerve fibers and blood vessels. Bone hurts when it is injured. It bleeds when it is broken. It is constantly being built up and broken down. Also, in spite of its popularity as a shade of wall paint, "bone white" is not the hue of living bone. Imagine instead a pale rose.

As part of our skeletal system, bones, in the simplest sense, keep us from being a puddle on the floor. But they also serve as cages, boxes, and basins to protect vital structures and act as incubators for new red blood cells. We are born with more than a hundred more bones than we have as adults, for many of our bones fuse together in the first years of life, such as the bones of the skull, and a final few, the bones of the pelvic girdle, don't fuse till puberty.

There are exactly 206 bones in the adult human skeleton, and they range in size from smaller than a tooth (inner ear bones) to bigger than a forearm (the femur, the longest, largest, and heaviest bone in the body). Likewise, bone names run the gamut, from dull (the frontal bone, the nasal bone) to surprisingly clever. The hip bone known as the *innominate,* for instance, literally means the unnamed bone, "so called," Henry Gray explains in *Gray's Anatomy,* "from bearing no resemblance to any known object." Many bone names come from Greek or Latin roots and are evocative even before tracing their literal meaning. When I first heard the names of the bones of the wrist, I thought of planets, a whole new solar system: *scaphoid, lunate, triquetrum, pisiform, trapezoid, trapezium, capitate, hamate.* And, in fact, the lunate bone is so named for being moon-shaped.

Whether delicate, dislocated, fractured, shattered, arthritic, or amputated, bones figure prominently in the daily life of the average physical therapist. Consequently, osteology, the branch of anatomy focusing on the bones, has been a large part of this course, especially in lectures. In lab, we always dissect down *to* the bone but not necessarily into it, as our primary objective is to examine in context the muscle, tissue, tendons, and ligaments that may sustain damage when bones are diseased or injured. We have become well acquainted with many parts of the skeleton by this point, just a couple of weeks shy of the final. But by lab's end today, we will uncover twenty-seven more bones, all of which can neatly fit inside a glove, and do: the bones of the hand.

Anyone peeking through the lab door might easily mistake the lot of us for the world's most serious manicurists. Two students hover over each hand at each body, cutting and tweezing small strips of skin. I am paired with Rachel, while on the other side of the table, Becky and Jenny handle the left appendage. These are a man's hands, big and meaty.

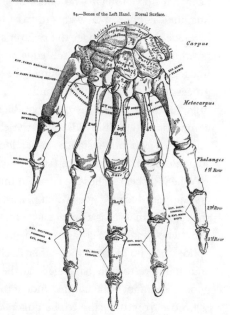

I have taken the place of Liz, the self-described surfer chick from Santa Cruz, who dropped out of school just yesterday. I get the feeling I make a better Liz than Liz did. According to Rachel, she had found this class " 'too sciency,' " the very thing I love about it, and Rachel seems awfully glad to have a new partner. She welcomed me this morning by putting a fresh blade on a scalpel for me.

One might think that

skin removal would be old hat by now. It is not. The process is tedious but at the same time allows the opportunity to marvel. As the
point of contact between our insides and the outside world, the skin,
the body's largest organ, differs dramatically depending on what it is
covering. Consider the back of the hand, for instance. In fact, consider your own. The skin here is loose and supple, to allow for the
tightest balling of the fist. When the fingers are relaxed, you can pull
up folds of the eyelid-thin skin, something you cannot do anywhere
on the reverse side of the hand, the palmar surface, where the skin is
not only thicker but tightly enmeshed with the underlying fascia.
This fixed flesh makes it possible to grip without the skin's sliding in
place. The sole of the foot is likewise fixed, so you literally don't slip
on your own skin, while the skin on top is loose.

Once the back of our hand is stripped, Rachel and I try supinat-
ing the arm so that we can move on to the palm, but we don't get
very far. Rigor mortis and embalming have made the limb so stiff,
we're afraid we might break a bone in the forearm. Not wanting to
hear that horrible *snap*—no doubt deserving of several demerits, I
imagine—neither of us wants to twist too hard. Using gentle pressure, we slowly maneuver the cadaver's arm into a fairly workable
position, jutting just over the edge of the table. I brace it while
Rachel begins removing the skin of the palm. This is a much slower
process than the back of the hand but also more conducive to chat-
ting. To my surprise, I discover that Rachel and I have something in
common: we are "the elders" of the class. At thirty-four, though,
Rachel, an avid long-distance runner, looks a decade younger. She is
so petite, she has to use a step stool to work on the hand.

When Rachel reaches the crease at midpalm—the fold palm read-
ers call your lifeline—we switch places. Continuing with her story,
she explains that she is a former accountant, a CPA with a downtown
firm, who's making a complete career change. As she says this, two
options pop up in my mind: either Rachel was too good at her job
and therefore too unchallenged, or the opposite. "I wasn't a very
good accountant," she admits, ending the debate.

I move from the palm to the fingers. Hardest of all to remove, I find, is the patch of skin containing the fingerprints, which almost seems epoxied in place. It is as though the body refuses to give up these marks of identity. But finally, the scalpel wins out.

Denuded, our hand is a handsome specimen, carpeted in plush, pinkish fascia. I now use the back of a probe to separate the trio of muscles nestled in the ball of the thumb. This plump part of the palm is known as the thenar eminence, or, in palmistry, as the Mount of Venus. Two muscles run on top and another beneath, this deepest one being one of the most important in the entire human body, the opponens pollicis, the muscle that makes the opposable thumb possible. It is shaped like a small feather and is the distinctive anatomical feature that allowed our primate ancestors to handle tools and manipulate their environment in ways other mammals could not.

It gives one pause—*This slip of a muscle,* I think, *helped advance our species*—but just a pause. Then a snip.

I cut it in half to more readily find the tiny nerve that activates it, a small branching of the median nerve. Given time, Rachel and I could trace the median all the way to its point of origin, over an arm's length away, deep within the neck. But today, we focus on just this small section, often referred to by hand surgeons as the "million-dollar nerve." No, the dollar amount is not a tribute to the nerve's value but is more of a penalty. If it is accidentally severed during a procedure—a carpal tunnel release, for instance—the patient will no longer have a functioning opposable thumb. The million dollars is a likely starting point for a malpractice settlement. (Far more often, I should point out, it is the patient who accidentally damages this nerve, cutting it when slicing a bagel, say, or shucking an oyster.)

The hand is a minefield of nerves. Rachel adjusts the overhead lamp for better viewing. It is also crawling with veins and arteries and lumbricales, the long, winding muscles that help us stretch and extend the four fingers. (*Lumbrical* means "wormlike.") Like all muscles, lumbricales would be useless were it not for tendons, the fibrous cords that unite muscle with bone and, like the strings of a marionette, make movement happen. Certainly the most astonishing tendons we have uncovered are the long, slender pairs running

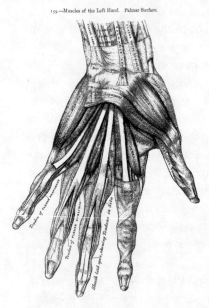

159.—Muscles of the Left Hand. Palmar Surface.

up each finger. They look like delicate reeds yet are obviously tough enough to last a lifetime—and beyond. They still work. By bending and unbending the cadaver's index finger, Rachel and I see one tendon sliding over the other.

The topmost takes its name, as tendons tend to, from the muscle from which it emerges, the flexor digitorum superficialis, located way up in the forearm. This muscle also has an old-fashioned name, which I personally think is a better name, one that had persisted for five centuries and found its way into *Gray's Anatomy:* the flexor digitorum *sublimis.* Why "sublime" rather than "superficial"? Well, purportedly, because the flexor digitorum sublimis sends its tendons to the fingers, including the single digit a romantic would consider the most important, the ring finger, and of course, marriage leads to sublime happiness, no?

As Rachel and I scrape off the last bits of fascia, she spills the remainder of her story: She's married. She and her husband have a home in the Oakland hills, two dogs—a nice life. "But I wasn't totally happy." She hated her job and really wanted to work with people rather than numbers. Years ago, a wonderful physical therapist had helped her recover from a horrific auto accident, and having been on the receiving end ultimately led her along the path toward today. But oh, that physical therapist neglected to say how much work becoming one would be! Rachel's midterm grades weren't so hot, she confides, and she is already stressing about the upcoming finals. "I have *got* to know these bones," she says, sounding like an accountant determined to make the numbers crunch.

"Well, let's make it happen."

I go to retrieve one of the five hanging skeletons positioned around the lab, to use for comparison. *I love these things,* I think as I wheel it to our table, how they are wired together like some crazy Calder mobile; how the bones rattle as you roll them; and, not least, how these are the real McCoy, not those bright white plastic clones made nowadays. Each skeleton is unique, different in size, coloring, and subtle surface markings, the bones themselves bearing a permanent imprint of the person who once was.

"I SEEM TO be always dreaming of writing or publishing some great fact—or facts—always hovering about, watching the adventurous ones and thinking of essaying to fly," yet at the same time, "very much disregarding the preparatory hops and short flights of those who now fly high," H. V. Carter begins his entry for March 18, 1854.

He could have stopped right there, as far as I am concerned. His words, at once somber and playful, soulful and angst-ridden, describe Carter so perfectly at this stage in his life that each time I read them, I find myself literally nodding in agreement. *Yes, this is the you I've come to know.* The wall between diarist and reader gives way, and I can see him as clearly as he is seeing himself. He sits in the Royal College of Surgeons library, his words suggest to me, taking refuge from the fumes of the laboratory and stealing a few minutes to write and think. Seated by an open window between banks of books, breathing in the fresh air, he watches people strolling in the park below, birds flitting about, carriages carting away the latest visiting Dr. So-and-So. Carter dips his pen again.

Over the past nine months, he writes, he has met here at the college many luminaries of the London medical world, many of whose works share space on the library's shelves, but the experience has not always been all he had hoped for. A person may think he knows a man "by his book, or published lectures, etc.," the twenty-two-year-old observes, but when you "see the man himself, hear him, speak to him, watch him," your opinion is "commonly somewhat shaken." In sum, "The whole resolves itself into this: The more one knows of a man, the more difficult it is to retain [a] favourable opinion, or implicit reliance on, him."

This seems to be one of those days when Carter can write no wrong, when he displays an insightfulness well beyond his years. The closing line of this short entry is also a beauty. "Two persons are generally concerned in every fact," Carter notes, "one discovers part, the other completes and corrects." The sentence has the pleasing concision of a maxim, one that could as easily apply to the two Henrys as it could to me and Carter, narrator and subject. Sometimes, though, I am baffled by facts he brings to light. Three months following this last entry, for instance, he makes a painful discovery in Henry Gray's just-published book on the spleen: he finds page after page of his artwork but no credit whatsoever for himself as the artist. Stunned, Carter takes to his diary. "See Gray's Book on Spleen," he jots. "Takes no notice of my assistance though [he] had voluntarily promised [to do so]. Rather feel it"—slighted, that is, and justifiably so. In a preface written expressly for the book, Gray had thanked by name the other colleagues from St. George's who had assisted him in various ways, warmly referring to each as a friend,

Illustration from Henry Gray's
On the Structure and Use of the Spleen

which surely must have made the omission sting all the more. Even so, Carter doesn't seem to hold a grudge for more than a sentence; immediately he goes on to praise Gray's book as "very creditable." And that is where it's left. He never confronts Gray about the matter, and Gray never brings it up.

At a loss to understand what happened, I find that I want to blame the publisher; it is possible a paragraph was cut from the preface, perhaps to save space, certainly without Gray's knowledge. To me, that is the only satisfying answer. Carter, I believe, had a very different explanation, one that reflected a significant change in his perception of the world. He thought he deserved it. The omission wasn't a sign from Gray. No, it was a sign from God.

Coloring his thinking was *Providence,* a Christian concept that colored my own upbringing. But Catholics do not spell the word with a capital *P,* as strict Evangelicals such as Carter did, and in that single letter lies a world of difference in how one views God's workings. I always understood providence in its broadest sense, as the belief that, while God has a master plan for all of creation, certain actions and events are beyond our comprehension. Though inexplicable, they are "providential," or, as God intended.

By contrast, Carter saw God as a hands-on manager of human affairs who, if pleased or displeased with an individual, intervenes through acts of Providence, sent from heaven like a personal bolt of lightning. This was a bedrock belief of Victorian-era Evangelicalism, as Cambridge historian Boyd Hilton has observed. God operated through a "system of rewards and punishments appropriate to good and bad behaviour," Hilton writes. "Almost always in the case of individuals, and sometimes in the case of communities [for example, an epidemic of cholera], suffering was the logical consequence of specifically bad behaviour. It could therefore incite as well as guide men to virtuous conduct in the future, but they must of course take the opportunity to examine their own actions in the light of their suffering."

Providence was clearly not a notion Carter invoked lightly. The word itself did not make a major appearance in his diary until he had flunked the M.B.'s in November 1853, his first unmitigated personal disaster. A week later, having picked himself up off the floor, he found himself "willing to view the matter as somewhat <u>Providential</u> in its occurrence," a diagnosis he had come to by a process of elimination. His "mind" was not at fault for the failing—he was certainly smart enough to have passed the exam, in other words—so his behavior must be to blame. Lax in his study habits, envious of others, neglectful of his religion, Carter had given God a host of reasons to intercede. Through the exam, He was sending H.V. a clear message. But in a perverse way, this slap from the hand of God ended up doing the young man some good. It made him *feel* God's presence. In failure, H. V. Carter found faith.

Like a scientist not wishing to publish a finding too early, he remained circumspect about his breakthrough. "Would not presume, but trust Christian principles [are] beginning to have some influence on [my] Character—specially that of Faith," he confides to *Reflections* in mid-May 1854. "Truly the very simplicity [of Faith] is almost a stumbling block," he adds, a line that comes across as a stunner for the reversal it shows in his thinking. What he had pursued so doggedly in the past had perhaps never really been so far out of reach. But then, as if the skeptic in him cannot refrain from questioning, he wonders: "Are we all together, body and soul, so much under God's direct influence and knowledge, that it may be said, the very hairs of our heads are numbered?"

The spleen book omission would be God's second warning to Carter. In his growing awareness that a Providence is never "any excuse for inexcusable deficiencies" but rather an impetus for soul-searching, he sets aside his disappointment and commits himself to succeeding on his second try at the M.B. exams.

And he earns his reward. "Success, under Providence," he writes on November 14, 1854, "M.B.—first division."

Hooray! he might have added, but he saves his effusiveness for a letter to Lily. "The only thing after all that can be said about the examination is that it is the <u>best</u> in the country. To have passed it implies a certain amount of knowledge." But more important, Carter must have felt he'd passed the greater test put to him by God. With a note of pomp, he then explains to Lily that degrees such as the M.B. are merely "<u>weapons</u> with which to fight the <u>battle</u> of <u>life,</u> and some men will fight as well with few as many. The contest will shortly begin downright, though, for many reasons, I do not greatly fear."

That Carter believed he had achieved a milestone in his spiritual growth is underscored just days later in *Reflections,* though it's what he *doesn't* say that is so telling. He stops writing here. His four-year-long running debate on whether to lead a Christian Life has quietly ended in God's favor. From now on, he folds his religious ruminations into his daily diary. However, none of this is meant to suggest that faith had come with inner peace. On the contrary, Carter remains as tortured as ever, albeit with an important difference. Whereas in the past he had agonized over the absence of God, now he agonizes over His presence. Sadly, though, he has little skill at recognizing God's signals. When your life is a lightning storm, as Carter's so often was, how do you know which bolt is a Providence?

At the start of December, Carter receives a request from John Sawyer, a figure who has been absent of late. Sawyer asks his former apprentice to fill in for him for a couple of days. For the younger doctor, this comes as a welcome chance to practice medicine and is also an unexpected compliment. Carter had covered for Dr. Sawyer for a full week back in September, a trial by fire as a G.P. that, frankly, had left him somewhat singed at the edges. All started well enough, he had written. Carter felt honored just for being asked (plus, he relished the chance to take time off from the college), but he ran into trouble almost as soon as Dr. Sawyer was out the door. The trouble being with himself. "Rather young appearance somewhat against me," he found, and I expect this was a fair assessment. The patients were used to the avuncular John Sawyer, after all, and here was this, this—

And who might you *be?* I can imagine being the first question to Carter from Sir Gordon Drummond, a regular of Dr. Sawyer's and a man whose name alone spells dyspepsia.

But it wasn't just his boyish looks that gave Carter away as green.

"Try too much to make self agreeable, which people who are ill don't so much relish as a more suppressed and sober manner," he writes one evening, diagnosing his own performance. Carter, who often lacked a certain social polish even in casual settings, found himself flustered and forgetful at those very moments when he could least afford to be—while seeing patients. "Omit many questions, etc., absolutely necessary to a careful enquiry into nature of 'case,' " he pointed out. But what made the young doctor most anxious was writing prescriptions. To his dismay, he didn't "find knowledge so ready at one's finger's ends" (or *fingertips,* as one would say today) and sometimes remembered the very best drug or dosage only after the patient had gone.

His week did end on a successful note, Carter happily reported, when he earned a fee of £1 for setting a broken leg. The patient had been a referral and wasn't a person but a bird—one Lady M.'s "favorite bullfinch!" Adept at taking animals apart, Carter was no doubt also gifted at fitting them back together. At least this is the impression one gets from his diary, which sometimes reads as if he were being trained as a veterinarian. At the college, an entire menagerie comes under his capable scalpel, from a walrus to a dog to a horse to a cuttlefish. Sometimes, too, he would also draw the creatures. In the summer of 1854, he had spent days dissecting and making drawings of a South American giant anteater, a magnificent creature he had actually gone to see when it was a popular (albeit short-lived) attraction at the zoo in Regent's Park. Given the task of anatomizing the anteater, Carter put up with fumes so foul he suffered from diarrhea and headache and ultimately earned high regard from superiors for his "artistic skill and praiseworthy industry." In spite of this, one might reasonably question the real-world relevance of the entire exercise, given that the only example of this peculiar animal in all of Great Britain was now dead.

Carter himself recognized the often esoteric nature of his work. "I don't dislike the occupation," he tells Lily. "In fact it suits me very well, though how it will further my practical experience, or enable me to cure more patients, I cannot tell."

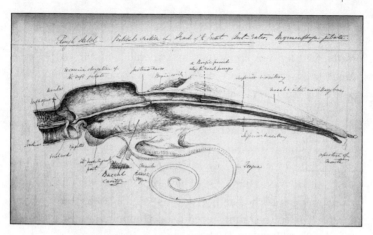

"Rough sketch" of an anteater, H. V. Carter, 1854

His restlessness with the studentship comes through most strongly with the start of the new year, 1855. At this point, with less than six months left at the college, he has little of significance to occupy him. Mr. Queckett is frequently ill and often not at work. Also, the junior student (to Carter's senior) had recently resigned and taken a "better appointment," a surgery with—as irony would have it—the General Screw Steamship Company, "a berth I once sought." But if hearing the younger man's news had hurt, Carter does not let on. He has both larger and smaller concerns. As he writes in early February, "The point is still, <u>what</u> to work at?" While he still entertains "dreams of [having a] delightful country practice," he is leaning more toward a career in medical research. His recent purchase of a high-powered microscope has opened new possibilities. "Have now sufficient confidence to trust [my] own powers in any branch of investigation concerning <u>human</u> anatomy"—note the distinction—"and now seem to wait for [the] opportunity."

The day after he turns twenty-four, opportunity knocks: "Professor Hewett made an offer that [I] should attend to [St. George's] Anatomy Museum and be a Demonstrator for £50 per annum—a kind and pleasing thing." If he were to take the job, Carter would primarily assist Henry Gray, in both the museum and the classroom.

"The offer is very tempting and in most respects very advanta-geous in its consequences." Still, there is some fine print to consider. Hewett had let him know that, as a member of the staff, Carter would be expected "to 'do something' to advance the reputation of the school," such as publishing a major scientific paper.

"And here's the rub: Gray has done very much, [he represents] a difficult precedent." Carter fears he will never measure up. What's more, he already imagines that people are insinuating, " 'Carter, look at him,' " meaning, *Look at what Gray has accomplished,* and by this point, it was quite a list. The twenty-eight-year-old multiple prizewinning published author, distinguished anatomy lecturer, and museum curator had most recently been appointed surgeon to the St. George's and St. James's dispensary.

Of course, Carter's concerns have less to do with Henry Gray, per se, than with how he sees himself. As he confides to his diary dur-ing a low moment, "Feel that [I] am *not* fit to ascend the ladder of distinction—must hold it for others."

But Professor Hewett sees none of this. *For heaven's sake, take the job!* he had urged him, even pointing out that the job offer might it-self be a Providence. "He reminded me of this," Carter writes. "I felt quite abashed and do now. Is this prospect from God?"

Henry Gray is equally encouraging. "A conversation with Gray has relieved some anxiety," he reports on his third day of wavering. Yet it takes another reassuring visit to Gray before he makes a final decision. "In short, the offer will be accepted, though with anxious feelings."

After all the hand-wringing, he makes a seamless transition from student to staff member, though the job does get off to a quiet start. He is a demonstrator without students to demonstrate to until win-ter session begins. When Henry Gray commissions him to do a se-ries of drawings, Carter doesn't hesitate. The subject: microscopic views of bone tissue.

As always, drawing remains perhaps the only aspect of his life over which Carter does not agonize. He purchases a lamp so that he can work in the evenings with his microscope, and he completes all forty illustrations for Gray in about as many days. "Gray called," he writes on October 7, "and liked very much [my] microscopic drawings." So much so, apparently, that several weeks later he approaches Carter with a far bigger project, though its scope doesn't seem to faze him in the least. "Little to record," Carter reports nonchalantly on November 25, 1855. "Gray made proposal to assist [him] by drawings and in bringing out a Manual of Anatomy for students: a good idea." (This "Manual" is the book that would come to be known as *Gray's Anatomy*. But that's jumping far ahead. At this point, Carter hasn't the foggiest idea of what fate has in store.) "Did not come to any plan," he adds, noting only that, for this project, he would not simply be the artist. He and Gray would be collaborating on the dissections.

Two weeks go by before Carter brings up the subject again, this time in oddly formal language, even for someone as formal as Carter. It reads as if he were submitting a petition for God's approval rather than writing a diary entry: "An employment has been offered by one of the Lecturers, a young man whose character as a man of science furnishes all needful example, which will both be profitable for body and mind and enable me to exercise the power of drawing which is perhaps my main and best standing point." Terms of the arrangement had been discussed. "Mr. Gray has acted fairly," Carter reports, "and probably £150 might be gained within the next 15 months."

Still, he has yet to accept the offer, and not because of nerves. Rather, he was praying about it every night, seeking God's guidance, waiting for a sign. And then, just before Christmas, he realizes that the sign is already before him—the proposal itself.

"Renewed conversation with Gray as to the proposed 'Manual of Anatomy,' which am to illustrate," H. V. Carter records on December 22, 1855. "May end in something. Gray shrewd, but considerate. The proposal seems Providential."

Ten

H ENRY GRAY CALLED IT THE *MENTAL PROCESS.*

Modern-day anatomists call it the *mental protuberance.*

Either way, I call it confusing. *What does the bump of bone at the base of the chin have to do with using your mind?*

I ask Dr. Topp one day as she passes by our dissecting table, and she answers the question with a visual. "Think of Rodin's *The Thinker*," she says, clearly trying to be helpful, though it seems she's just adding feathers to my pillow fight.

She then adopts the classic pose of deep thought—chin resting on fist—and the air is immediately clear. *Of course.* But then—

"Well, that's not literally why." She explains that *mental* actually comes from *mentum,* the Latin word for "chin," but by this point, the mnemonic has stuck.

"Now you'll never forget it."

I have come to love this quality about Dr. Topp—Kim, as she has said we can call her. She is pro-mnemonics, pro-etymologizing, pro-whatever-it-takes to get you to remember the correct anatomical term, which I consider a very Henry Gray–like trait. He often supplies in his text the derivation or translation of terms as well as memorable visual descriptions. At the same time, though, this makes Kim the polar opposite of Dana, who is anti-mnemonics. Too often students use them as a crutch, as a shortcut that bypasses true understanding, Dana believes. But the two instructors are of one mind when it comes to the end result. If the phrase "**T**om, **D**ick, **A**nd **N**ervous **H**arry" helps you remember the **T**ibialis posterior muscle, the flexor **D**igitorum longus muscle, the posterior tibial **A**rtery, the tibial **N**erve, and the flexor **H**allucis longus muscle—the anatomical mouthful that converges in the lower leg—that's fine, but don't stop there. You have also got to be able to identify these structures in a body, know their function, their point of origin, and so on. As for myself, once learning becomes knowing, mnemonics become super-fluous, I have found, and at the end of the day, nothing makes know-ing come faster than doing a dissection yourself.

I make the first cut for the four of us, a shallow incision from the top of the forehead down to just below the Adam's apple. In a sense, this is like the imaginary line drawn down the center of a room that siblings must share—Becky and Jenny have their side, and Rachel and I, ours—but there is no rivalry here. Indeed, we share a bond forged by the intensity of what lies ahead: taking apart the face.

Becky and I wield the scalpels for our teams. We both start at the same place, the skin at the cleft of the chin, and cut in opposite direc-tions. The underlying bone here is the mental process. About an inch below each of the lower canines (the pointiest teeth of the lower jaw) is a tiny hole—the mental foramen—through which passes a tiny nerve, the mental nerve. We burrow for our respective holes. The nerve at each site supplies the lower jaw and lip, and when anes-thetized during a dental procedure, for instance, causes localized sen-sory loss. Becky finds hers first, though I am soon behind. Each looks like a threaded needle, a small white fiber poking through a tiny eye.

The next procedure is more delicate: exposing the facial artery,

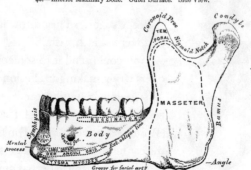

48.—Inferior Maxillary Bone. Outer Surface. Side View.

the major blood vessel supplying the face. Just as Henry Gray describes, "This vessel, both in the neck and on the face, is remarkably tortuous," though, for the dissector, "torturous" could apply as well. An offshoot of the carotid, the facial artery starts in the neck, curves over the mandible, gives off branches to the lips and nose, then terminates at the inner "canthus" (from the Greek *kanthos,* meaning the corner of the eye). For Becky and me both, dissecting its winding branches takes considerable time and care. But once finished, what is revealed is far more than the twisting vessels themselves. Her side and my side mirror each other, an unexpected and remarkable display of the body's inner symmetry.

Next, we go our separate ways. This is the last lab of the course and, for the students, a chance to review areas of difficulty before the final exam. Rachel and Jenny head to the feet and Becky focuses on the brachial plexus. For me, this is a last chance to explore. I decide to examine the TMJ (the temporomandibular joint, also known as the jaw joint), a dissection we'd not had time to perform.

I immediately hit an obstacle: the ear.

The lab manual contains no instructions on how to remove an ear, so I operate on instinct only. I pull the pinna away from the scalp and slice in a slow circle. I dig at a sharp angle as if cutting a weed at its root. This proves startlingly effective. The ear comes off in one piece, and now I face a dilemma I had not anticipated: *What does one do with an ear?*

288.—The Pinna or Auricle. Outer Surface.

Becky suggests I just throw it away, but that seems a little rash.

"You could send it to someone," Jenny offers wryly, making an allusion to Vincent van Gogh, who put his ear in an envelope and gave it to a prostitute for safekeeping. *(Wasn't her name Rachel?)* For now, I fold the ear inside a towel and set it aside.

Where the ear had been is now a hole in

the head the size and shape of a kidney bean. This is the entrance to the ear canal, which runs just posterior to the TMJ. *No wonder my clicking jaw sounds so loud,* I say to myself. For many years, I have had what's officially called TMJ disorder—TMJ, for short—a catchall phrase for various problems at this juncture of two bones. Mine, exacerbated by teeth grinding while I sleep, mainly manifests as a *pop* or *click* whenever I yawn or chew or open my mouth too far. In my head, the sound is huge. I am always surprised people don't hear it.

Having TMJ has made me hyperaware of my jaw, which, for the first time in my life, I am finding to be an advantage. The clicking is like a TMJ sonar pinging in my cheek, guiding my hand as I dissect. I cut back the parotid duct and gland, then gently slice through the powerful muscles of mastication. Next, alternately feeling with my fingers and using both the blade and the butt of my scalpel, I peel off the layers of fascia covering the joint itself. This is like peeling the skin from several stubborn cloves of garlic—tedious work, but I am utterly absorbed. I am a *ping* hunting a *pop* to its source.

After almost two hours, I have exposed a near flawless specimen of the temporomandibular joint, including one of its most delicate features: the tiny cartilaginous disk that acts as a kind of shock absorber between the temporal (upper) and mandible (lower) bones. When this disk is damaged or, as in my case, abraded, TMJ disorder results.

Stepping back from the lab table to appraise my work, I have no qualms about praising it aloud: "That is beautiful." My lab partners heartily agree, as does Kim, who moves about the room telling other groups to come see my dissection. I am doing a final cleanup of the sur-

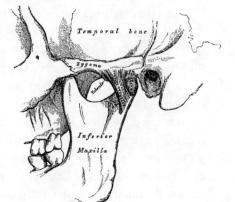

104.—Temporo-Maxillary Articulation. External View.

rounding tissue when I see over my shoulder a wash of inquisitive faces. They have been quietly looking on, as if I were a sculptor brushing the dust from my latest work and they did not want to disturb me.

"Nice job," Casey remarks, breaking the silence. "How'd you do that?"

With that, I become the go-to guy for the TMJ. Three groups invite me to their tables, where I tug at ears and talk through the procedure.

By the time lab is over, I feel as though I have truly graduated from this course in anatomy. Even so, I am not content. Too many gaps in my knowledge—*mental foramens* of the metaphorical kind, to coin a phrase—remain to be filled. I have never studied the brain, for instance, and I am still trying to work my way through the nervous system. Kim suggests that I take one last anatomy course—this one, for medical students. She promises to e-mail me the details as well as an invite to a postfinals celebration party she would be throwing at her home in a couple of weeks.

NEARLY EVERY STUDENT comes, a testament to the fact that I was not alone in the affection I had formed for Kim. The sole absentee, Rachel, had opted to take the course as an incomplete rather than risk getting a failing grade. She will have to retake PT anatomy next summer. Becky and Jenny got As, though, as did most of my earlier table mates. As it turns out, I had gotten a passing grade of my own: Kim had used my TMJ dissection in the lab portion of the final, where test cadavers have numbers pinned to specific parts. *Name the muscle at pin A and describe its role in the movement of the TMJ.*

"You saved me from doing that dissection myself," Kim says, clinking a Snapple bottle to my Heineken. "So, thank you, Bill!

"Come on," she adds, "it's time for a tradition." Sam has set up a digital camera on the deck railing for a class photo in the backyard. He sets the automatic shutter, then hops down and squeezes in next to Kim and me.

Click.

In the moment, I cannot help but flash on the photograph that launched me on this journey six months ago—Henry Gray and his students in the St. George's dissecting room. I look at that collection of faces several times a day, as I have made the photo the wallpaper on my computer. It never fails to seize my attention. I no longer zero in only on Henry Gray but notice other characters: the gentleman seated directly to Gray's left, for instance, who, with his mutton-chops and formal black coat, looks like the heir to a cough-drop fortune; and, just behind him, the younger fellow, arms crossed at the wrist, who appears to have lost his left hand. Then there are the two scamps at the very back who stand on either side of a human skeleton. Is the one on the left holding the skeleton's hand? But what I mostly see in this picture now is a missing person—H. V. Carter.

By the time this photo was taken, March 1860, Carter was no longer the demonstrator of anatomy. He was no longer even affiliated with St. George's Hospital or, for that matter, a resident of London. In fact, he would not come to call England his home again for thirty years.

What had happened in the interim? Where had Carter disappeared to and why? Well, the full, tortuous story comes complete with a torrid scandal straight out of a Victorian novel. But first things first. The two Henrys still have a masterpiece to create. And just two weeks into their collaboration, Henry Vandyke Carter has hit a major obstacle.

Eleven

"[H]AD A] LONG CHAT WITH GRAY, WHO CANNOT UNDERSTAND that anyone should really wish to work and yet not be able to begin," Carter confides to his diary on January 8, 1856. "He is altogether practical—'Do it!'—his aim 'money,' chiefly. As for self, need energy and right counsel. Mind certainly not healthy or balanced, and time very indifferently spent."

This sounds less like a chat and more like a spat. Every time I read this entry, I feel as if I am right *there* as these two enact a fascinating early scene in the genesis of *Gray's Anatomy*. Each Henry plays his role to a T: Gray, the taskmaster, is all business, while Carter is the temperamental artiste who seems to have misplaced his muse. Oh, he sounds so grieved, so misunderstood! But he also sounds troubled. The line *Mind certainly not healthy or balanced* always stops me cold. Though I can make no claims to a definitive diagnosis, this young man seems gripped not just by artist's block but by a debilitating depression as well.

That H. V. Carter was prone to dark moods had been clear from the start of his diary. To some extent, I had taken this tendency with a grain of salt, as I know that sometimes, on the pages of one's diary, feelings get overblown, the better to puncture and purge them. Also, I had noticed a predictable pattern to Carter's moodiness. He suffered from a condition I had myself as a young man, what I call the Sunday syndrome. His entries tended to be at their longest, most heartfelt, and most angst-ridden on Sundays, the day when he set aside worldly matters and took time to reflect and attend church. The sermons delivered by the ever-stalwart Reverend Martin were rarely less than "capital" or "excellent" and always left Carter with a boost of fresh resolve to be a good, moral, industrious person. But this was the spiritual equivalent of a sugar high. By the end of the day—diary-writing time—he would crash and burn, convinced that he fell far short of the ideal Christian he knew he should be. Such was his misery that I've often thought he would have been a happier person if the week had only six days.

As the darkness slipped into his diary more and more often, I could no longer write it off to Sunday. With the start of 1856, Carter's moodiness becomes the blackest melancholy, and the twenty-five-year-old writes of being fitful, fatigued, and overtaken by lethargy, classic physical symptoms of depression. Carter knows he is not well but is at a loss for what to do. "Am right down [sic] helpless when ought to help self. No abiding effort. Must be <u>compelled,</u> not invited."

Remarkably, he pushes through this latest episode, but it takes him a full four weeks. Finally he reports, "Made first drawings for the work."

Now just 360 more to go.

Gray's workload was no less daunting. He would have to write nothing short of an encyclopedia on anatomy in less than a year and a half. Under the circumstances, I would almost expect more quarrels and creative differences to have surfaced between the two men. But such did not seem to be the case. From the outset, author and artist shared a strong vision of the book they wanted to create. As historian Ruth Richardson observes in her introduction to the thirty-ninth British edition of *Gray's Anatomy*, "Neither was interested in producing a pretty book, or an expensive one. Their purpose was to supply an affordable, accurate teaching aid for students like their own."

As both men dealt with students on a daily basis and had recently been students themselves, they knew that small innovations would make a big impact. Unlike Quain's *Elements of Anatomy,* which came in a three-volume set, for instance, this book would contain in a single volume everything a student needed to know about the human body. Further, bucking the trend of pocket-sized texts, some as small as 4 by 6 inches (10 by 17 centimeters), Gray, Carter, and their pub-lisher, John Parker & Son, also planned a larger than usual book, with a no-squinting-necessary text size and illustrations that could breathe. Even at this size, 6 by 9½ inches (15 by 24 centimeters), it would still be light and easily totable. Parker & Son would also be happy, as this was a cost-effective size to print. The bottom line was, the book was shrewdly designed from the get-go to *sell*.

I cannot help bringing up a small irony here. The very phrase that H. V. Carter had used as an epithet to describe Gray also serves as a perfect characterization for the book the two had in mind: it would be *altogether practical*. What's more, practicality would be a guiding principle throughout the project's eighteen-month duration. Be-tween author and artist, there would be no wasted effort. Perform-ing dissections together, for instance, would save time on many fronts, including helping them come to a speedy agreement on the fine points of each illustration—what stage of a dissection should be drawn, what perspective to use, and so on. As seasoned anatomists, too, they certainly knew how to make the most of their most pre-cious resource, cadavers. Between dissections done for classes and those for the book, no material would go to waste. I expect the same could be said when it came to the manuscript. Gray undoubtedly drew upon his three-year back catalog of lecture notes as a basis for his text and, alternatively, used any freshly written text as a basis for new lectures. This, I believe, helps explain the distinctive tone of Gray's prose. You can open the book to almost any paragraph and find the clear, unrushed voice of an experienced instructor speaking directly to a rapt classroom.

Carter was able to do double duty as well, using dissections he had performed as demonstrator or for the Anatomy Museum as subjects to be drawn for the book. At first, he drew on paper, but about six months into the project, he made a radical change. He began draw-ing directly onto the wood blocks that would ultimately be used for the book's engravings. Whether at the publisher's behest or, as I sus-pect, on his own initiative, this would end up saving a huge amount of time by bypassing the need to have someone else transfer the drawings from paper to wood. Still, it was akin to switching to a whole new medium, and Carter found the transition bumpy. "Pretty assiduous at Kinnerton Street," he notes after one of many long days spent in the lab. "Drawing from nature and on wood. Result of latter, so-so. Require practice, improving." Strikingly, he does not sound frustrated or discouraged, which is right in character, since Carter was always happiest when engaged in learning something new. And one need look no farther than the finished book to see how perfectly he mastered the technique.

I should mention one last major time-saving device Carter employed, one that, upon first learning of it, came as a surprise to me: he copied some of the illustrations from other anatomy books. This fact, omitted from the later American editions of *Gray's Anatomy*, was acknowledged right up front in the original English edition at the beginning of a seven-page list of illustrations (also left out of later editions, if only to save space). The number of copied illustrations was small, 77 of the 363 total, and it is easy to understand Gray and Carter's rationale: If another artist has perfectly captured a dissection, why not use what's ready-made? You would save not just time but a cadaver, which, in the spirit of the endeavor, seems eminently practical. But what merited borrowing?

As it turns out, Carter pulled not from one or two but from nineteen different sources, including his beloved Quain's. This discovery invokes the wonderful image of the two men raiding the St. George's Lending Library and of Henry Gray's home office being carpeted with dozens of anatomy books laid open to possible candidates.

I TAKE A seat as Ms. Wheat studies my list of the nineteen artist-anatomists. It literally goes from A to Z—*Arnold*, first name *Friedrich*, to *Zinn*, first name *Johann*—and comprises a who's who of leading figures of the nineteenth century, not only English anatomists but German, Italian, French, Scottish, and Dutch as well. Most of these luminaries have since faded into obscurity, however, and copies of their works are now exceedingly rare. Which is the reason for my visit.

Ms. Wheat hands back the list and asks just one question: Where would I like to begin?

"With Arnold," I reply without hesitation, and not simply because he is alphabetically first on the list. A third of Carter's copied illustrations come from this single source. Friedrich Arnold (1803–90) was a longtime professor of anatomy at the University of Heidelberg, and he specialized in the microanatomy of the nervous system. He was the author of some sixteen books, most of which he also illustrated, one of which is now being delivered to me personally.

Ms. Wheat places a foam lectern in front of me, followed by *Icones Nervorum Capitis* (1834), Arnold's first illustrated work, a monograph on the cranial nerves. It takes just a preliminary fanning of the pages to understand why Carter would wish to copy from it. Arnold's artwork is seriously beautiful, as is the book as a whole. The pages are oversized and the lithography of the highest quality.

I linger over one of Arnold's full-page drawings, a profile of a human head split down the middle—a hemihead. Though the line work is astonishingly precise in spots, the overall effect is sumptuous, almost painterly. Arnold's style is unlike any I have ever seen, perhaps even deserving its own category. Call it Anatomical Romanticism.

With just a turn of the page, though, I begin to see a flaw in his approach to illustrating, one that Gray and Carter no doubt noticed as well. Friedrich Arnold produced illustrations in pairs, the first being a fine-art rendering; the second, a simple bold outline, as if for an anatomy coloring book. Only on this second page are the names of parts listed, so you have to flip back and forth between the two prints to get the full effect. While this would be only a minor nuisance for a student—as when footnotes, for example, are at the back of a book rather than at the bottom of the page—it was still a format that Gray and Carter would not wish to duplicate. When drawing from Arnold, therefore, Carter had to perform an act of imaginative superimposition.

I pull out my copy of *Gray's* and look for a good example of this melding and find one in Carter's illustration of the fifth cranial nerve, taken from Arnold's paired illustration of the same subject. Here, the artistic and the diagrammatic combine seamlessly, with Carter's added innovation of the anatomical names appearing on the parts themselves. Seeing the versions side by side also helps clear up a bit of confu-

257.—Distribution of the Second and Third Divisions of the Fifth Nerve and Sub-maxillary Ganglion.

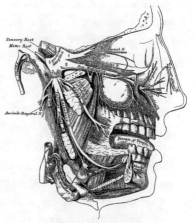

sion I had been carrying with me. In his characteristically meticulous way, Carter identified three degrees of borrowing in the list of illustrations, noting that drawings were either directly copied from another source, "Altered from," or "After," distinctions that sounded to me like shades of gray on the same cloud. But I finally understand what he meant. Carter categorized this particular piece, for instance, as "After Arnold," which sounds right. It is not a line-for-line copy, nor is it altered from the original to, say, highlight a different anatomical feature or aspect of a dissection. No, instead, it is a tribute to a great German artist-anatomist. It is an homage.

Ms. Wheat, in her mysterious librarian stealth mode, has quietly come and gone, leaving behind a neat stack of three: Quain's *Anatomy.* The work of English anatomist Jones Quain (1796–1865), this was Carter's second highest source for borrowed images. I approach it with trepidation—no, make that fear. To my dismay, I had discovered that one of my favorite H. V. Carter drawings, the glorious full-page engraving of the muscles of the back (reproduced in chapter 6), was not an original but copied from Quain. Was his version little more than a Victorian Xerox, or did he bring something of his own style to the reproduction? Did he make it *his*?

I find the answer in volume 3.

Except for being about 50 percent smaller, the engraving at first glance looks almost identical to Carter's. But something is different, too, something it takes me a moment to appreciate. Simply put, Carter's drawing, by comparison, seems to lift off the page. He employed every tool at his command to create three-dimensionality, from a greater variation in line width to an off-center light source. Rather than each layer of the dissection's looking uniform, as in Quain's, Carter "lit" his version to create a subtle play of shadows across the subject's back. Also, Carter had taken Quain's original squat figure and stretched it, making the torso taller and slimmer, thereby accentuating the illusion of depth. In spite of these adjustments, he indicated that this drawing was "directly copied from" Quain, but in the copying, the twenty-five-year-old still brought his own aesthetic to the piece.

Comparing other Quain drawings to Carter's, I find the same pattern again and again. It's as though Jones Quain created a first draft of each, which H. V. Carter then polished and perfected.

I tell Ms. Wheat that her seemingly magical ability to retreat through the rear door of the Rare Books Room and return with the most esoteric of tomes makes me wonder if she could produce the impossible: the greatest anatomy book *never* published. What I'm referring to is a legendary nonbook by none other than Leonardo da Vinci (1452–1519). As the story—a true one—goes, Leonardo first began giving serious thought to producing a book on human anatomy while in his midthirties. He thought he would call it *On the Human Figure*. Along with the title, Leonardo jotted down a rough outline and did some early sketches. Frankly, though, this was a pretty big idea for someone whose knowledge of human anatomy was quite small. At this point in his life, Leonardo's anatomical education had come chiefly from reading outdated texts, such as the works of Galen, Mondino, and Avicenna, and from the observation of surface anatomy in living models. His exposure to human dissection was limited to being a spectator to the occasional postmortems open to the public. This changed once Leonardo moved from Milan to Florence in the early 1500s. He was now able to obtain the random arm or leg of unclaimed corpses from a Florence hospital and, working by candlelight in the hospital basement, surreptitiously began teaching himself internal anatomy. As his understanding grew, his conception of the anatomy book changed as well, moving from being artistic in tone to something far more scientific. A *Treatise on Anatomy*, he retitled it. Still, the book remained a perpetual work in progress, set aside again and again while he and his restless mind pursued other projects, most of which he also never finished.

Possibly the best chance the book had of seeing the light of day came in the year 1510, when Leonardo met a potential collaborator, a highly accomplished young anatomy professor named Marcantonio della Torre. According to a sixteenth-century source who'd heard the details secondhand, the two men agreed to join forces and had divvied up the responsibilities. Marcantonio would organize Leonardo's extensive but scattered notes and write the text while Leonardo would create the illustrations. Now, whether this collaboration ever really existed remains a matter of debate among Vincian scholars. Leonardo himself never mentions it in his notebooks. Had it come to pass, however, the two might have become the Gray and Carter of the Italian High Renaissance. But Marcantonio died of the plague in 1511, and that was the end of that.

The anatomy notebooks, like the painting of the *Mona Lisa*, remained in Leonardo's possession for the rest of his life, and few people ever got a peek at them. Following Leonardo's death in 1519 at age sixty-seven, the collection became the property of his companion (and possible lover) of the previous dozen years, a young man of twenty-six named Francesco Melzi. Francesco stored the trove at his family's villa near Milan, where it went virtually untouched for fifty years.

Enter the bad heir.

Upon Francesco's death in 1570, his nephew inherited the collection and, soon after, allowed the anatomy notebooks to be broken apart and sold off. Over time, an untold number of drawings was lost, some probably destroyed for their heretical nature, but, sometime in the early seventeenth century (the date is undocumented), some somehow became the property of the Royal Library at Windsor, England. Access, however, was limited, if not impossible. It is believed that Charles I himself locked the Leonardo papers in a large chest within the library, where they remained, like objects in a forgotten time capsule, for well over a hundred years.

Then came the hero of this story, Robert Dalton, and what must have been the most extraordinary day of his career as royal librarian.

The exact date is not known, only the year, 1760; George III had just taken the throne. The key to the chest had long been lost by this point, but that did not stop Mr. Dalton from finding some way to pry it open. Perhaps curiosity had simply gotten the best of him, or more likely, he had been roused by the mention of Leonardo's name in an old library inventory list. Either way, could Mr. Dalton ever have been prepared for what he discovered? At the bottom of the chest was a stack of miracles on paper, 779 drawings in all.

Alerting the king to this great find was *not,* my gut tells me, Mr. Dalton's very next move. He surely allowed himself the indulgence of examining each and every drawing. Who could resist? Plus, wouldn't a careful study fall under his purview as royal librarian? His Majesty would expect nothing less than a full accounting of the collection.

Once informed, George III invited Britain's leading anatomist, William Hunter, to inspect the drawings. Dr. Hunter, who, along with his brother John, operated an anatomy school right next door to St. George's Hospital, reportedly told his students that Leonardo's work was three hundred years ahead of its time. "I expected to see little more than such designs in Anatomy as might be useful to a painter in his own profession," Dr. Hunter admitted. But he instead saw the disciplined work of a "deep student," adding, "When I consider what pains he has taken upon every part of the body, [and] the superiority of his universal genius, . . . I am fully persuaded that Leonardo was the best Anatomist, at his time, in the world."

Even so, his work was essentially kept a well-guarded secret for another hundred years; Gray and Carter, for instance, certainly never saw and perhaps never even heard about the notebook pages.

Finally, in the late nineteenth century, some of Leonardo's drawings were exhibited in public and published in book form to great acclaim. As for Leonardo's *Treatise on Anatomy*, however, it remained untouched, unread, unseen, a glorious volume on an imaginary bookshelf of what might have been.

MS. WHEAT SLIPS back into the room bearing what looks like an oversized shirt box but of expensive design. She sets it down with a gentle thump and says with a sly smile, "You'll need to wear gloves for this." She pats the box with both hands.

"Could you clear some space?" she asks as she moves to the desk drawer where she keeps the hand wear. One of Ms. Wheat's least glamorous job responsibilities, it turns out, is to take home the used pairs of reader's gloves and wash them with her own fine washables. "Oh, you get a fresh pair!" she notes, not the "underwear gloves," as she calls the others. Indeed, this pair is ultrabright and new. I pull them on and carefully unwind the string that secures the top flap of the box. By this point, I have read the label on the side of the container, but I am still amazed by what I find inside: an original copy of the greatest anatomy book of all time, Andreas Vesalius's Renaissance-era masterwork, *De Humani Corporis Fabrica* (On the Structure of the Body).

"I thought you might want to see this," says Ms. Wheat, barely concealing her own delight as she leans over my shoulder. What lies before us is the second edition of Vesalius's book, dated 1555, produced a dozen years after the first and considered the definitive edition.

Though I have never seen a copy, much less touched one, I feel as though I know this book cover to cover, so pivotal is its role in the history of anatomy and also, it is no exaggeration to say, in the history of Western civilization and culture. The *Fabrica* is an atlas of the human body based on actual dissection of the human body, a practice rarely performed and widely reviled at the time it was written. This alone would have made the book significant. But, as Vesalius made clear, the book also came with a radical agenda: to dismantle the mighty doctrine of Galenism, a vast body of knowledge based on the writings of the second-century Greek physician Galen. Though riddled with errors and anachronisms, Galen's work was still considered sacrosanct and Galen himself a revered figure. The man was to medicine what Jesus Christ was to Christianity; to challenge Galen's word was nothing short of heresy. Vesalius knew he would be stepping onto a minefield in publishing his book, yet his ambition was matched by his fearlessness, which was equal to his canniness. He was not going to change fourteen hundred years of entrenched thinking with a slapped-together pamphlet, he realized. No, he had to create a volume that was perfect in its packaging, inside and out. Which is exactly what he did. Proof lay on the table before me.

This copy of the *Fabrica,* Ms. Wheat explains, was donated to the library by a successful anesthesiologist who dabbled in antiquarian books. He believed it had been "locked up in a monastery for four centuries," she notes, and, by the looks of it, I would say he was right. The cover is absolutely unmarred.

"And look"—Ms. Wheat points to the edge opposite the spine—

"It still has its sheen," I say, completing the thought. Like the gilding on an old family Bible, the edges were finished with a fine red varnish that makes the book look fittingly grand.

Ms. Wheat leaves me alone with the *Fabrica* while she attends to another patron. I carefully turn the cover and page to what serves as the sixteenth-century equivalent of an author photo, an engraving of Andreas Vesalius at work in his laboratory. The woodcut shows a swarthy, bearded man with an uncompromising gaze. To his left is a partially dissected cadaver, suspended upright, which he is using to demonstrate the long tendons of the hand. Rather than wearing a dissector's smock, Vesalius is dressed like a prince in an ornate tunic, as if symbolically distancing himself from anatomy's unseemly reputation. The man is immediately fascinating.

Andreas Vesalius, born in Belgium in 1514, entered medical school at age nineteen. Fortuitously, this was right at the time when Galen's en-

tire oeuvre was being printed in its original Greek for the first time. Up until
this point, students had been studying
Galenism through translations of translations. Now it was as though a long-
neglected masterpiece were restored to
its original brilliance—at least, that's
how most scholars viewed the new
Galen. But not Vesalius. Fluent in classi-
cal Greek and able to read Galen's actual words, Vesalius saw the cracks and
began making notes. The first threads
of the *Fabrica* took form.

Andreas Vesalius

In the years following medical school,
Vesalius gradually gained experience in dissecting corpses, something Galen had never been able to do, as dissection was forbidden
in ancient Greek society. Vesalius attended autopsies and also managed to obtain bodies of executed criminals and unclaimed corpses
for his own private study. In this respect, he was very much like
Leonardo, though the world would not have to wait long to learn
his thoughts.

Like Henry Gray, Vesalius became a professor of surgery and
anatomy (at the University of Padua, Italy) and, in his every spare
moment, devoted himself to writing. He spent two years on the first
edition of the *Fabrica,* which he completed in 1542. Deliberately aiming for a lofty tone befitting a scholarly work, he wrote in a highly re-
fined form of Latin. While I unfortunately would not know high
Latin from low, I find myself enjoying the text for its visual aesthetics alone. Each page is exquisitely designed and composed. And the
woodcut illustrations, still celebrated for their fidelity to human
anatomy, are extraordinary. Well over four hundred fill the book.
Some are as small as thumbnails (including one of a thumbnail); others are elaborate two-page foldouts with detailed keys.

As with Gray, it is often assumed that Vesalius was his own artist
when in fact he worked closely with an illustrator (uncredited,

though most likely it was the Flemish artist Jan Stefan van Kalkar) and served as art director, to use a modern term. Nowhere is Vesalius's hand more evident than in the many drawings of full-length figures. In spite of how extensive each dissection may be, each figure is depicted in an active, lifelike pose. One skeleton looks like a musician playing an invisible saxophone; another, as if he's taken a break from orating to show off his abdominal viscera. With these startling images, Vesalius was attempting to humanize what many considered an inhuman practice, to show that anatomy is a science of the living body. As distinctive as the poses are the settings. The figures are often standing atop a mountain plateau, a kind of Vesalian Valhalla, in my view. Here again, Vesalius was making a powerful point visually—literally elevating the cadaver, the source of anatomical truth.

Illustration from the *Fabrica*
by Andreas Vesalius

Deliberately mirroring the structure of Galen's works on anatomy, Vesalius divided the *Fabrica* into seven "books," or sections. But instead of paying homage, he systematically laid out all of Galen's errors and corrected them one by one. No, the human liver does not have five lobes but two. (Galen had counted five in a dog and concluded people must also have that number.) No, arteries do not originate in the liver but in the heart. And no, once and for all, animal anatomy is not the same as human. Vesalius's brazenness infuriated old-school anatomists, including teachers who had once been his mentors, but Vesalius's work was too persuasive to be dismissed. And, in yet another canny move, he ensured that his ideas would be widely disseminated. Specifically for students, he produced a condensed and less expensive version of the *Fabrica* and also published a German version, as Germany at the time was a major publishing market. With surprising suddenness, dissection became de rigueur in medical schools, and Vesalius himself became something of a celebrity, attracting huge crowds to his lectures and dissection demonstrations. So highly esteemed was he that Emperor Charles V named Andreas Vesalius his personal physician. Today he is known by a grander title: the founder of the modern science of anatomy.

Speaking of which: I would love to spend more time with the *Fabrica*, but I have an anatomy class to attend. Pulling off my gloves, I am first puzzled, then slightly horrified to find that the fabric at the fingertips is dirty. Careful as I was, traces of the five hundred-year-old ink rubbed off. Or, to put it another way, traces of *Vesalius* rubbed off. Which almost makes me wish I had saved them. By the time this thought hits, though, it is too late. The cotton gloves are bound for Ms. Wheat's laundry basket, and I am up in the anatomy lab pulling on a pair of rubber ones.

AS THIS IS my third first day in just six months, one would think the scene around me would be utterly familiar. Well, yes and no. The freshly polished linoleum gleams. The blackboards have been sponged clean. The lab guides at each station are centered and new. But what transforms the lab into something completely different is the suffocating crowd. It is two minutes past the hour, and 143 first-year med students are scrambling to find their assigned tables among the cadavers, squeezed in three to a row, eight rows long, as eight instructors direct traffic.

Luckily, I arrived just before the big crush of students and easily found my table, number 24, back in the left north corner. This is a great spot, as it's right next to a bank of windows with a spectacular view of the Golden Gate Bridge. For the moment, there are just two of us here, me and a fresh cadaver—female, aged eighty-eight. Then, out of the sea of periwinkle and green scrubs, Dr. Topp appears with someone in tow. "Bill, this is Kolja; Kolja, Bill; you'll be lab partners, okay? Great!" she says, and then, as though taken by a rogue wave, she is gone.

"Kole-ya, is it?" I say, pronouncing it just as Kim had.

He nods and, as if anticipating the question everyone asks, explains that, yes, his parents named him after Kolja Krasotkin, the young hero in Dostoyevsky's *The Brothers Karamazov,* but no, he is not Russian. He adds that he lives in Berkeley and just finished a Ph.D. in chemistry at U.C. Berkeley.

Kolja looks ten years late to class. He is about thirty years old, I would guess, and wears sandals and an oversized T-shirt instead of the traditional sneakers and scrubs. His long blond hair is pulled into a loose ponytail. But what is most striking about Kolja is how mellow he seems. In fact, I cannot help asking, "So, *are* you a med student?"

"Oh, yeah," Kolja replies, "I just decided to head back to school." He's still looking for his niche, he adds.

Our four other table mates arrive all at once, and after quick introductions, we huddle around the lab manual. Our first act as a team is to split up. Every table of six is divided into twos, and each pair has a separate assignment for each hour of lab. We are to rotate at the hour and at the end of class come together and review. This tight schedule was born of necessity. Med students have so much to learn, so many subjects to cover in their first two years, that their entire curriculum is accelerated. For this anatomy course, three months of work is squeezed into six weeks. Today alone, they have to complete in three hours the equivalent of three separate labs.

First up for Alex and David is starting the dissection of the thorax, while for Marissa and Erica, it is a task just a few inches below: dissection of the abdomen. Kolja and I have been consigned to studying prosections set up at a long table in the center of the lab. Personally, I would rather dissect, but for someone like Kolja, this is not a bad way to begin. Prosections are like anatomical flash cards—good, fast learning aids. We will be able to examine quickly the anatomy that the remainder of our team is slowly cutting through. But there is a problem at the Island of Prosections. It has been overrun by students.

The sight brings to mind vultures swarming around a half-eaten corpse: the table is encircled by other students with this same assignment, all poking and probing the detached parts. Their bodies form a wall, and Kolja and I and a couple of dozen others are already shut out. No instructors are around, as they are all off helping with dissections.

"I think this is partly personality screening," Kolja comments as we wait for a spot to open up. "You know, weeding out anyone who's not an alpha male. A 'survival of the fittest' kind of thing—"

"Yeah? So how do you respond in situations like this?" I ask.

"Not well," Kolja admits, and indeed, he suddenly looks paralyzed.

With that, my inner alpha male ascends. "Okay, people," I say in a raised voice, "can we get in here, please? Please?" I tap someone on the shoulder—"Hey, can you scoot over a little?"—and she obliges. Kolja and I settle in. "Okay, let's get started."

Lying before us is not only a selection of preserved specimens—hearts, lungs, and rib cages—but also a fresh cadaver that Dana and Kim had dissected earlier. Its entire chest cavity has been reassembled so we can go through it part by part. It reminds me of a teaching tool popular during the seventeenth century, "Anatomical Venuses," as they were called—life-sized human models made of wax, with removable parts—except that this Venus has a penis.

Going from the outside in, I first show Kolja the two visible layers of skin, the epidermis and dermis, as well as the major chest muscles, pectoralis major and minor, parts that I had learned so well under Kim's guidance. Next, I remove the entire rib cage so we can examine the "intercostals," the space between the ribs. The three thin layers of intercostal muscles—external, internal, and innermost—are clearly distinguishable, thanks to Dana and Kim's expertise. What they have done is akin to exposing the different layers of a Triscuit, showing the various weaves of wheat. "See how the muscle fibers of each go in a different direction?" I point out.

"Cool," Kolja responds each time I introduce a new anatomical part.

Moving on, I show Kolja the internal thoracic artery and vein—"They don't run parallel to the ribs, it's important to remember, but lateral to the sternum"—and prod him to finger three portions of the parietal pleura (costal, diaphragmatic, and mediastinal), to help him remember how each is different. In this dissected cadaver, with rib cage, lungs and heart set to the side, it is easy to find the phrenic nerve, lying like a loose guitar string from the neck down to the diaphragm. "You can't breathe without this thing," I tell Kolja, then share the one mnemonic I had learned from Dana: *C-3, 4, and 5 keep the diaphragm alive.* "That'll be on a test, I can almost guarantee."

As we continue, a young man across the table interrupts: "Hey, can I ask you a question?"

"Me? Sure," I respond, "but I'm not a TA or a professor—"

"Well, are you knowledgeable?"

I don't hesitate. "Yes."

"What's this?" He points to a prominent vessel in the cadaver's neck.

"Well, first, tell me what you think it is," I say.

"The superior vena cava?"

Everyone else at the table is listening in by this point.

"No," I answer, "the superior vena cava heads directly into the heart. The one you're pointing to—see how it goes under the clavicle?—that's the subclavian vein."

"Oh, of course, the subclavian! And it changes names, right? Into the, what is it, the 'axillary' vein?—"

"Right, very good, it changes *from* the axillary *into* the subclavian once it hits the armpit, right at the level of the first rib." This guy is not just a good guesser, I've figured out. Like everyone here, he has taken basic anatomy before, but he also obviously came to lab super-prepared. Still, there is a big difference between book smarts and body smarts, and this particular vein is very confusing. "Okay now," I continue, "the subclavian changes names a third time, right? Do you know at what point?"

199.—The Axillary Artery, and its Branches.

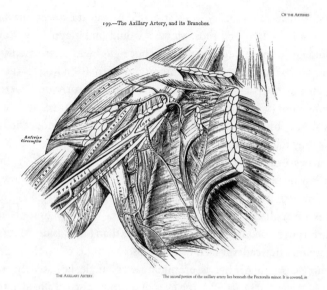

Anterior
Circumflex

THE AXILLARY ARTERY.   The second portion of the axillary artery lies beneath the Pectoralis minor. It is covered, in

"No idea."

"Right here"—I point to the spot—"where it joins the internal jugular and becomes the brachiocephalic, which feeds right into the, the . . . ?"

"Superior vena cava!"

"Perfect," I say, only now realizing what is going on here: I am actually teaching anatomy to these students (med students, no less), something I would never have imagined myself doing six months ago.

At 2:30, Kolja and I return to table 24, where we find Dana coaching David and Alex in the final phase of removing our cadaver's lungs. One moment the organs are inside the body, and the next, boom, boom, David and Alex are each cradling one, looking like the proud papas of lumpy gray twins. I congratulate the pair.

The same scene is playing out at tables all around us, and by now, the energy in the room is through the roof. It is hot and sweaty, even by the window, and as I take Erica's spot at the cadaver, I find that we are standing almost butt to butt with students at the table behind ours. If this were a reality show, it would be called *Extreme Anatomy* or *Speed Dissecting,* and most viewers would find it appalling. But I would not want to be anywhere else but right here. These med students are so quick and keen and hungry to learn. Even Kolja is catching up with the pack. He is a fearless dissector, as it turns out. For this second hour of lab, he and I are stationed at the abdomen, and Kolja has already uncovered two difficult-to-find arteries embedded in the dark, oily folds of the mesentery.

Inches away, Erica and Marissa are working feverishly on the thorax, having picked up where David and Alex left off (they are now at the prosection table). At one point, Kolja and I stop to watch as Erica slices through the pericardium and Marissa carefully uproots the heart from its bedding in the chest. Rather than put it aside, Marissa, aware that our cadaver died of heart disease, holds on to the organ, turning it over in her hands, examining every angle. I can tell that she is looking at the heart with the eyes of a future doctor, a healer, trying to see where the organ broke.

This had definitely been a momentous day for her, Marissa tells me later in the afternoon as we are cleaning up. "I'd taken an anatomy class but never done dissection before," she goes on to explain. "I thought I was going to hate it, but"—she stops and, as if confiding a secret, adds in a whisper—"I loved it. I totally loved it." Marissa looks out the window for a moment. "It's funny, I came into school thinking I was going to go into pediatrics, but now I don't know. I . . . I think I might like to do cardiology."

Out in the hall, I chat with other students and repeatedly hear the same refrain: they liked dissecting, they liked it a *lot,* which is something I do not remember ever hearing said with such enthusiasm in the other anatomy classes. But then, of course, these are budding doctors. The *body*—in all its fleshiness, complexity, gruesomeness, and beauty—speaks to them, *sings* to them, in a distinct and powerful way. My favorite remark comes from Rayuna, an Indian woman with the carriage of a ballet dancer, who tells me with infectious delight: "It's really *crowded* in there," referring not to the anatomy lab but to the abdominal cavity. "Amazing, just amazing, how everything twists and turns and wraps around." She pauses, as if picturing a whole body in her mind. "It was so cool to see inside."

At least one student hadn't been quite so wowed. Blake, whom I find sitting on the floor slumped against his hall locker, looks totally drained by the experience. After a friendly hello, I ask what had surprised him most about his first day of lab.

"That my cadaver was a woman," he replies without hesitation.

Of all the things I might have expected him to say, that was nowhere on the list.

"I assumed I'd get a man," Blake clarifies. "I don't know why, I just thought I would—so I was really . . . *surprised* that I got a woman."

He must have seen the *huh?* still stuck to my face. Blake confides that his grandmother is terminally ill, and he'd dreaded the idea of having to dissect an elderly female cadaver. Then he got one, and, wouldn't you know it, he discovers that the cause of death was the same lung condition his grandmother is suffering from, COPD, chronic obstructive pulmonary disease. "That was kind of upsetting."

Blake looks down at his hands for a second. "But then, it was kind of weird," he says. "I had to dissect the lung, of all things, that was my assignment, and I almost felt bad or guilty because I didn't freak out. I just, you know, followed the instructions and basically chopped it off and scooped it out with my hands—" He winces at his inelegant phrasing. "I almost wish—"

"You'd had a harder time doing it," I say.

"Yes." He falls quiet.

"Well, hey, look at it this way," I tell him. "It's your first day of medical school, and you've already learned what must be one of the hardest lessons."

He gives me a skeptical look. "Yeah? What's that?"

"Keeping your emotions in check so you can do your job."

Blake manages a half smile and nods.

Twelve

H. V. CARTER WAS ONLY TWENTY AND NOT YET A DOCTOR WHEN he began doctoring his own mother. This went way beyond making pills for her, as he had in the past. He now felt competent and confident enough to help direct her care. During his time home for Christmas break 1851, he gave his mother an extensive physical exam, after which he wrote up his findings and recapped her entire medical history in his diary (on Christmas Eve, of all nights). What had heretofore gone unsaid he now makes clear: his mother, forty-one years old, had breast cancer.

"M." had discovered a lump fourteen years earlier, Carter writes, but, "suffering no inconvenience," paid it no mind until seven years later, when the mass began to grow and the pain became chronic. Even so, it would be three more years before M. traveled to London to see a specialist—Henry Gray's mentor, the famed Benjamin Brodie, as it turned out—and her illness was finally diagnosed.

Now, Carter reports, "Whole size of mass equal to palm of hand nearly," the pain has spread to her hips, and she is taking a long list of medications, the dosages for which he carefully notes. While a local physician visited her regularly, Carter had been closely monitoring his mother's treatment over the past two years and, in a sense, following her case since he was a boy. He was just thirteen when she first fell seriously ill, and sixteen when, in the summer of 1847, she was diagnosed. At that very time, it is fascinating to note, Carter had just started his apprenticeship in a Scarborough medical practice, and by year's end, he would be in London to attend the same medical school where—yes—Dr. Brodie was serving as professor emeritus. Now here's where coincidence stops being coincidence, I believe, as mother and son's histories merge. Though some historians have surmised that Henry Vandyke Carter chose medicine over the family business, art, because of the influence of a science-minded uncle, I think otherwise. It surely had to do with his mother. Perhaps if he became a doctor, the teenager must have thought, he could save her.

Given the gravity of her condition at Christmastime 1851, it is surprising to find Eliza Carter continuing to appear in her son's diary. The mentions, year after year, are typically brief and blunt—"M. worse," or "M. weaker"—like addendums to her case notes. So much so, in fact, that I started wondering if H. V. Carter ever saw her as more than just Patient M. I know of only one way to find out. I make an online visit to the Wellcome Library's Carter catalog and do some shopping.

Just twelve letters from H.V. to his mother have survived, surely a small fraction of the number he wrote over the years, but from these, a clear and striking impression emerges. Yes, she is very much a patient in her son's eyes, but she also plays a key role as his spiritual confidante, the only person with whom he can be nakedly honest about his struggles with faith. "Prayer is the one [subject] I have had least resort to, hardly ever," he admits in one letter. "The Bible read daily, but how? Not prayerfully. This subject is so discouraging. You must know, dear Mother, such a turmoil within me, is very unfavourable."

Frankly, I would have expected Carter to keep this turmoil to himself, for fear not only of burdening his frail mother but also of disappointing her. But the reality is, he felt he had no one else to turn to. Lily, Joe, and their father were nowhere near as devout as he, and, even after eight years in London, he still had not a single friend in whom he could confide. "Where shall I look for a Christian friend?" he laments again and again in his diary, most recently in February 1856. Sadly, the person with whom he spends the most time, Henry Gray, does not qualify, though Carter can see God in their relationship. "It does seem the act of a kind Providence to have brought me so much into contact with such a character," he notes in June of the same year. Still, he yearns for "inter-communion" with like-minded others, he adds in the same entry. "I have literally none." Even his longtime pastor does not seem very sympathetic. On two separate occasions, Reverend Martin had advised him not to write down his religious struggles—advice that, needless to say, Carter did not heed.

He poured out his troubled soul to his mother, and his letters are undeniably moving. Yet it is also hard not to see the two as tragic kindred spirits, she as her health is failing, he as his faith is failing. Just as he hoped to save her, she now wishes to save him. So close is their bond that mother and son create a private ritual. On appointed nights starting at 11:00 P.M., H.V. (in London) and his mother (in Scarborough) engage in an hour of "mutual prayer." By joining together, they are amplifying their appeal to God and reinforcing "the spiritual state dear to us both," as he describes it. But on an earthly level, they are also communing with each other in an intensely intimate way.

For all his candor, Carter is surprisingly silent on certain topics, which frustrates my inner eavesdropper. I want to hear about experiences he does not record in his daily diary. Two of the twelve letters, for instance, were written while Carter worked on *Gray's Anatomy,* and I would have loved a bit of news on his progress, or even just a quick postscript about his artistic life. Unfortunately, he didn't write a word; fortunately, there *was* a firsthand witness.

"Henry and I have a very (too) quiet life at present. He is at home drawing, etc., a good deal, and I am out 'studying' pictures a good deal," Joe Carter reports in one of a handful of letters to Lily Carter from this period. "We generally are both home in the evening: reading, smoking (not I, yet), drawing, retiring (and reappearing next morning) rather late."

The Carter brothers had moved here to 33 Ebury Street, their second London apartment together, in mid-August 1855. Joe had failed to get accepted into the Royal Academy—much to both brothers' disappointment—so he was schooling himself, spending his days among the old masters at art museums. But he was not anywhere near the student H.V. had been. I get the feeling, in fact, that going out and "studying pictures" was a euphemism for roaming the galleries and eyeing the young ladies. What's more, the anatomy lessons H.V. had given his brother apparently had not stuck. "Joe diligent," Carter told his diary, "but progress very slow. <u>Cannot</u> draw figure." Nevertheless, Joe, a watercolorist who, like his father, favored landscapes, certainly knew how to set a scene with just a few bold strokes, as demonstrated in a letter to Lily written not long after the boys had finished unpacking.

"We are becoming used to our new quarters," Joe writes, adding, "of course H. is the principle [*sic*] <u>decider.</u>" Fortunately, this apartment has "two windows and two doors and two cupboards," Joe notes playfully—as if saying, one for each of them—and it came with a sofa and an easy chair. The latter "has attracted the favourable notice of Harry," Joe explained (apparently using a family nickname for H.V.), and the former "is dedicated to"—that is, piled high with— "drawings, folios, and other <u>untidy</u> objects."

Joe adds more details in a later letter: H.V. uses the main room for drawing, he notes, and also for seeing "his visitors," a reference to the medical students Carter had begun tutoring. As for himself, Joe boasts, "I have got a 'bona fide' attic upstairs for my studio," evoking an image of a garret lit with candles and with pinned sketches papering the walls. Had he not been sharing the apartment with his persnickety, abstemious Christian brother, his life in London would almost sound Bohemian.

By this point, Joe, who would turn twenty-two in December 1856, had definitely begun taking himself more seriously as an artist (lest there be any doubt, he signed his letters to Lily "*J. N. Carter, Artist,*" as if imitating the signature of his idol, the English painter J.M.W. Turner). But the fact is, he was not an inspired one. I have seen some of Joe's paintings; at best, they look inspired by his father. Ironically, the truly gifted artist was not up in the attic studio but down in the main room, drawing anatomy on wood and smoking late into the night.

H. V. Carter would never have seen himself this way, nor would he have viewed his art as Art. Art was framed and hung on a wall and admired. His work for *Gray's Anatomy* was scientific and academic, chiefly, and, by its very nature, too morbid to be displayed or even discussed in polite company (which might explain why Carter didn't write to his mother about it). His drawings were meant to benefit the student, not to bear evidence of his hand. Even so, even without a tiny *H.V.C.* in the corner, his style is so distinctive that I, for one, can easily tell a Henry Vandyke Carter drawing from that of an imitator.

Joe Carter also had a blind spot when it came to his own gifts, I believe. He describes himself to Lily as an "indifferent correspondent" and apologizes for his careless writing, but he was quite mistaken. He was a wonderful, evocative writer, much more so than H.V., who, despite keeping a diary for many years and writing hundreds of letters, did not have Joe's ease of language. Lily must have loved getting her little brother's letters. They are full of clever wordplay and fresh observations—they are, in a word, charming, as Joe must surely have been. For instance, he opened one letter to her with a lovely riff about the persistent nature of one's own history: "It often surprises me to find how intimately the past becomes interwoven with the present, and the apparent future," he begins. "And I have, at times, immensely wondered to find that what is past—the past—does not, nor will it, detach itself and remain where it was (or where it might have been intended to have remained) but it must bring itself forward, and smilingly, or otherwise, present itself as an old friend, and will not be denied. It is not till we try to remove or change old ideas or facts that we find how deeply rooted they are."

OVERNIGHT, THE FUTURE has arrived in the dissection lab: eight sleek new computers were installed on the north side of the room for the purpose of playing CD-ROMs of virtual dissections. One of them is stationed right next to our table. The CD-ROMs are an adjunct to the students' studies and, incidentally, something to keep them occupied while waiting for a spot at the prosections table. Nevertheless, the presence of computers in the lab signals a momentous shift. This is where the study of human anatomy is headed, some experts say, to 3-D re-creations and simulations that do away with cadavers entirely.

Until then, there is still "cadaver splatter" to worry about, not to mention gunky hands. Hence, the computer keyboards and mouses are covered in Saran wrap; high tech meets low tech. There are also skeptics to convert, such as Dana Rohde. As she points out, "Why sit and watch a video or CD-ROM when you can just go dissect?"

Truth be told, Dana is not a big fan of prosections either. "Most of them are awful. They're old, they're dried out, and they've been handled by so many people." Worse, prosections present in pieces what should be taught as a whole. "You simply can't learn that way."

Dana does not mince words, even between bites of a vegetarian Subway sandwich. She and I were sitting outside the Health Sciences building, between classes, on a gorgeous September afternoon. We had gotten together to catch up belatedly on our respective summer adventures—she in the Galapagos Islands with her twin sister; me, in the PT course—but talk had quickly turned to the anatomy program. Dana explained that the course I am attending is actually quite different from the one taught just a few years ago. Up until the year 2000, first-year med students at UCSF took six full *months* of anatomy, which was pretty much the standard for medical schools across the country. "Only four students per cadaver, and they dissected literally everything, from eyeballs to brains, genitals, toes. Everything."

This "old curriculum," as Dana called it, was indeed old, harking back to the 1830s, when legal cadavers started becoming widely available due to a change in law, first in England and, soon after, the United States. As a result, dissections by students themselves (not just by instructors and demonstrators) were feasible. The half-year-long anatomy courses that Gray and Carter took as students and taught as teachers became the norm, and, in fact, those classes were not substantially different from the ones offered 150 years later. Every other class in a modern med student's curriculum had changed, however. For instance, Gray and Carter never had to study radiology, oncology, and immunology, nor genetics and molecular biology, the fields that have revolutionized medicine in the past fifty years. By the late twentieth century, the typical four-year med school curriculum had become so jam-packed that, short of adding another year, some courses had to be scaled back. To many administrators, the traditional six months of anatomy was starting to look like a luxury, particularly given the huge costs involved not only in acquiring and maintaining cadavers but also in staffing. As I had come to appreciate firsthand, having up to eight instructors supervising fledgling dissectors several times a week certainly must not be cost-effective.

In 2001, UCSF became one of the first medical schools in the nation to make a major move, implementing a change so radical as to cause an uproar from the students. The school eliminated the traditional anatomy course; integrated into other courses a fraction of what had formerly been taught (the anatomy of the heart and lungs, for instance, was taught in a class on the organs); and dispensed entirely with cadavers and dissecting by students. The small amount of anatomy still in the curriculum was taught with prosections. As UCSF is one of the top-ranked schools in the country, other med schools soon followed its example and started slashing their anatomy programs.

While it was an academic year Dana would rather forget, she also takes pleasure in recounting how a great many students successfully lobbied for the reinstatement of the course (albeit reduced from six months to the current six weeks, supplemented by some anatomy classes spread throughout the year) and the reenlistment of cadavers. Even so, repercussions of that failed experiment remain, as I would soon witness.

Following lunch, I accompany Dana to an appointment in the dissection lab. She is meeting with two fourth-year students, a young man and woman who had been part of the test class of 2001. We find them at the back of the empty lab peering into a nightmare cookie jar, a human head with the skullcap removed. They had contacted Dana because they were about to start their ophthalmology rotation and were worried about gaps in their knowledge.

The pair bump heads over the head as Dana takes them on a quick behind-the-eyes tour, pointing out how tight the packaging is. "Now you can see why a pituitary tumor *here* gives you an optic nerve

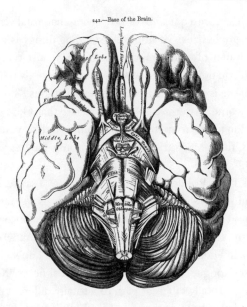

242.—Base of the Brain.

problem *here*"—the students are nodding, clearly getting it—"and why a carotid aneurysm at that level gives you a cranial nerve VI injury, which affects . . ." She studies their faces, waiting for a response, waiting, waiting—

"Lateral eye movement?" I offer after a long moment.

"Exactly," Dana says. "Very good."

I step back while Dana finishes up with the fourth-years. To their credit, it strikes me, these two young people knew they were missing something and wanted a remedy. But how do you know you don't know something? And what about all those students who are not here in the lab?

As Dana and I return to her office, I ask, "Are you worried that these students with less anatomy training than in the past will be ill prepared as doctors?"

"Ultimately, they'll be fine," she says without hesitation. "They'll know enough."

I'm not satisfied with this answer, as Dana can tell.

"I think it's more a question of not having that 'total vision' of the body," she emphasizes, "of not understanding things as well as they could. So much of understanding anatomy is just tying it all together, and you don't get that when you do little body parts." At the same time, she is not unrealistic in her expectations. "I don't expect them to become anatomists. No, I appreciate that I know the body so well, I don't have to memorize anything. And, as you know, my big thing is, the more you understand anatomy, the less you have to memorize."

"Yes, Dr. Dana Rohde, the anti-mnemonicist," I say teasingly.

She laughs, then slips back into teacher mode as we stand alone in the hallway: "Take the cranial nerves, for example. Once you've dissected them, you can picture cranial nerve VII coming out of the brain stem and going through the skull, and you know exactly how it gets to the tongue. And likewise with cranial nerve IX—you just see it taking a totally different path to a totally different part of the tongue. And you'd never even *think* you'd have to memorize it." You would simply *see* it, she reiterates. "That's the vision I have."

WE HAVE SOMETHING that the other students are lining up to see: we have a good cadaver. Actually, that's understating it some. A good cadaver is one in which the structures come clean easily, separate distinctly, and are not surrounded by excessive amounts of fat or obscured by calcification. What makes our body not just good but *very* good (and very popular) is that it has a fully intact reproductive system. Given that donor cadavers are generally quite elderly (the average age in this group is eighty-four) and, if female, have typically had hysterectomies, this is a rare sight. Certainly, it is my first.

Our cadaver's uterus is about the size of a fist, lavender-colored, and supple to the touch, unlike in the prosection, by comparison, where the organ has literally shriveled to the size and texture of a walnut. What's more, while the prosection is missing the Fallopian tubes, here they are in perfect shape, extending from the uterus in twin arcs. (Fallopian tubes are not attached to the ovaries.) At the tips of each tube are the tiny egg-grabbing fingers called fimbriae and, just below these, the ovaries, plump and almond-shaped. The stabilizing ligaments and surrounding tissue are likewise in place, clearly visible through the glossy peritoneum draped over the uterus and ovaries. In fact, the whole looks in such good working order, even in an eighty-eight-year-old body, that it is strangely easy to imagine, to *see*, the system in operation: An egg being pitched from an ovary. The fimbria, hovering overhead, catching it in its grasp. The mesosalpinx flapping gently, nudging the egg through the Fallopian tube to the uterus, where it unites with a sperm cell. And finally, in time-lapse motion, the uterus expanding, filling with life as if filling with breath.

Once you have images like this banked in your head, you cannot help viewing people's bodies differently—*anatomically*. You see life with a kind of picture-in-picture feature, I have discovered. Your friend breast-feeding her newborn becomes an astonishing multiplex image, a body feeding a body it has created. The jogger running down your block is a churning red machine. The vision works just as well on yourself, turning even the most prosaic of actions golden. My morning pee, for instance, will never be the same.

The urge-to-go that gets me out of bed now comes with its own series of illustrations. In my mind, I can see the bladder, a small, delicate organ, stretched to capacity, like a balloon that won't

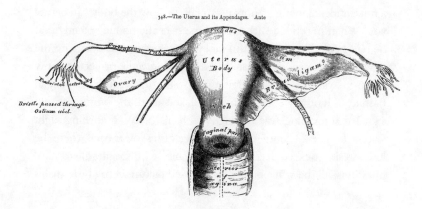

348.—The Uterus and its Appendages. Ante

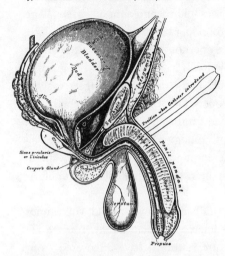

341.—Vertical Section of Bladder, Penis, and Urethra.

survive one more blow. I can picture it right above the chestnut-sized prostate gland and pressing against the thin muscles of the lower abdomen, making the surface of my belly feel as taut as a snare drum. In the moment before splash-down, I know that the visceral afferent nerves in my bladder are flaring, sending distress signals to my brain via the spinal cord: *Empty me now!* Once the keg is tapped, so to speak, and the pressure reduced some, a larger picture starts coming into focus. From the bladder, I can mentally trace the twin tubes of the ureters crossing the pelvic brim and ascending in graceful lines up to the kidneys. Within each kidney, I see inside the complex filtration system that has strained this thin pale yellow stream from my blood. And by the time I flush, I have glimpsed the greater complex of blood vessels leading back to, and out of, the heart.

Having a vision of how the body works also comes, naturally, with a finer understanding of how it can fail, of how the body can betray you. When my friend Richard told me over the phone recently that he had been diagnosed with kidney cancer, it was as if, before the news sank in, a slide carousel had dropped into the projector in my head: views of a kidney—anterior, posterior, hemisected—began flashing behind my eyes. As Richard talked about the symptoms that signaled something was not right with his body—drenching night sweats, fatigue—I zoomed in on the area of his lower back where the kidneys sit. I pushed deeper, peppering him with specific anatomical questions, all the while building a detailed picture in my head of his diseased organ: Which kidney—right or left? (*Left.*) Where was the tumor? (*Right on the surface; two inches in diameter.*) Had it penetrated the bedding of fat? (*Yes.*) What about the renal hilum? (*No.*)

"Well, that's good news," I said, visualizing how, when caught at this early stage, such a cancer probably could not spread.

"Yeah, it is," Richard answered. "On the list of cancers to have, they say this is not a bad one. I didn't have to have chemo or radiation. They simply removed it—" At this, I could imagine the procedure—the renal artery and vein being transected, the kidney and surrounding fat lifted out—yet I also found myself thinking, *What happens to the ureter? It must be removed, too, tied off at the bladder.*

The conversation gave me a small sense of the diagnostic skill doctors in training must develop, the ability to play out possible treatment scenarios in their mind. A doctor's vision is not always an enviable one, though, as Meri, a fellow student in the anatomy lab, helped me appreciate. One afternoon she told me about a friend of hers, a recent med school graduate, whose mother had developed a life-threatening autoimmune disease. "She *knows* how bad it is, what's happening inside her mom," Meri said. "But she really can't *tell* her mom everything—it's all too awful. And when her mom asks her questions about her condition, sometimes she just doesn't answer. She doesn't want to tell her mom what she knows."

June 20, 1856
33 Ebury Street
London

Dearest 'Ma,'

I beg of you to send for me, if you feel the least inclined, or if you think me capable of doing any good, however little. This brings me at once to a little request of yours: can I propose any remedy (marvelous! it must be) to restore you at once to health and strength? Oh! ho, dear Ma, you don't ask this, do you? It is but a little thing. . . .

If Carter sounds desperate at first, he had good reason. He had just learned that his mother's health had taken a serious turn. But I find it all desperate, even as he tries for levity. In his dissembling, his utter helplessness is all the more palpable. Turning to his diary soon after, he reveals the depth of his fears: not only is his mother's health failing, but now his grandfather is "dangerously ill." He worries, "Are these shadows of coming events?"

Yes, they are. Within days, the grandfather dies and Carter returns home for the funeral. "Find M. certainly changed," he reports. "She really does look in a sinking state—very pale and thin, with an anxious expression." In the entry, Carter also notes that he has prescribed opium as well as "Quinine Chloric Ether," an anesthetic, which suggests his mother was in considerable pain.

After this, M. all but disappears from the diary. Then comes this:

Sun., April 5, 1857

> *This evening, at about 9:00, the landlady brought up a Telegraphic Dispatch, which contained the words, "Your mother died this morning. Come on Tuesday if you can, not later than Wednesday."*

Arriving home, H.V. and Joe find their mother laid out in her bedroom covered with a death shroud she had made for herself several years earlier.

Eliza Caroline Carter was forty-six years old.

In her final days, she did not send for her doctor son. *"Not that I should not be delighted to see him,"* his mother had said, a friend of the family told H.V., *"but he will sit and watch me so earnestly, and can do no good."*

Thirteen

I LOOK UP FROM THE BODY AND FIND THE LAB EMPTY SAVE FOR
Anne, an assistant who has been prepping dissections for the next
day's class, but she's on her way out. "Turn off the lights when you
go," she calls, and the door springs shut behind her. It is only six
o'clock but feels much later. The black October sky has turned the
bank of windows into a mirror. I see myself and the class cadavers.
All but mine are zipped up for the night in their white body bags.

I like being in the lab at this hour. There is a quality to the silence
that reminds me of the libraries I loved as a child. My mind quiets as
I focus on the task at hand, which tonight involves finishing up the
day's last assignment, a complicated dissection of the anterior thigh.
I am doing this to help my lab partners, true, but also for my own ed-
ification. Since yesterday, we have been engaged in a three-day
"Limb Lab," an extensive exploration of the arms and legs, which includes studying a part of the body most people don't even know exists: fascia.

Before I started studying anatomy, I certainly had no idea that we
have under our skin a kind of second skin. And in truth, over the
course of the courses, I have really viewed fascia only as the tissue
one has to cut through to get to the "good stuff." Henry Gray was
never so dismissive. In fact, I now think of him as a passionate pro-
fascia-ist. He considered fascia no less important than muscles in the
overall composition of the body and gave the two equal billing in the
third chapter of *Gray's Anatomy*. Aware that this was an unconventional way to present the material, he felt the need to explain himself. The muscles and the fasciae are described conjointly, Gray
writes in his introduction, because of the "close connexion that ex-

ists" between them. With that point made, he adds an observation of another sort. "It is rare for the student of anatomy in this country to have the opportunity of dissecting the fascia separately." When one presented itself to me, therefore, I jumped at the chance.

And whom should I find but Professor Gray himself? There on page 11 of the lab manual was his classic description of fascia, almost like an epigraph introducing the dissection I had to complete. The wording was all quite technical, really—"*The fasciae are fibroareolar or aponeurotic laminae of variable thickness and strength found in all regions of the body,*" and so on—but I could not help noticing that one tiny clause had been lost in transcription. In the original, Gray notes that *fascia* is Latin for "bandage," a simple fact that conveys a helpful image. Like a bandage, fascia wraps around, covers, protects, and binds. There is no better place to see this bandaging effect than in the thigh, where the body's largest and longest muscles are found.

As a time-saver, deskinning of the upper thigh was supposed to have been done in advance by the teaching assistants, but they had obviously run out of time before getting to our table. I don't mind. I look upon it as a chance to pay extra-close attention to the layer of "superficial fascia" that undercoats the skin. In appearance, superficial fascia is as different from the second type of fascia, "deep fascia," as apples are to oranges; in fact, it brings to mind oranges, fittingly enough. If you could invert your skin, à la an orange rind turned inside out, you would find the entire surface lined with a similar soft, spongy whitish material. That is superficial fascia. It makes for a great insulator and acts as a support structure for sweat glands and superficial nerves and blood vessels. By contrast, deep fascia is more fibrous and therefore tougher than superficial fascia, and, as its name indicates, it is located deeper within the body. In the thigh, deep fascia is called fascia lata, so named, Gray explains, for its "great extent" (*lata* meaning "broad"), and, sure enough, I find this taut, opaque tissue wrapped around the full extent of the thigh like a big Ace bandage. It not only binds the thirteen meaty muscles within but also intensifies the force they generate.

To cut it apart, I choose scissors. With the tip, I puncture a hole in

the fascia lata at the lap line, then scissor downward to the middle of the kneecap. It is as easy
as cutting fabric. I make crosscuts at the very
top and very bottom and peel the long double
doorways back.

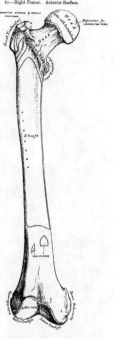

I have just cut my way into what's called
the anterior compartment of the thigh, the
largest of three fascia-walled sections surrounding the femur bone. Each compartment
is literally a discrete room housing a set of
muscles related to one another by function.
For instance, the thigh's main extensor muscles, which help extend the leg, are all bundled
together in the anterior compartment. My
next task is to unbundle them.

There is no better tool for this than the fingers. (Gloved fingers, mind you.) I press into
the mass of undifferentiated fibers, feeling for
seams. Deep fascia binds these individual mus-
cles together as well, but it's of a different consistency here—more
like a sticky fluid, viscous and clear—and I easily work the muscles
apart. The first to pull free from the pack is the sartorius, the longest
muscle in the body, which extends from the hipbone to the inside of
the knee. By running it between my thumb and fingers, up and
down, up and down, I am able to clean its whole length. But in the
pulling apart, I have broken the fascial bonds that give the muscle
stability and support. The sartorius now drapes across the thigh like
a sadly sagging sash.

Unbeknownst to most people, the human body has several biceps
(a single muscle with two parts, or heads); the biceps that bulges
when someone says "Make a muscle" is simply the most famous.
There are two triceps muscles, triple-headers. But there is only one
quadriceps, the quadriceps femoris (or, as it is commonly known,
"the quads"), and it forms the main bulk of the anterior compartment. As I separate and clean each of the four muscles, I find myself

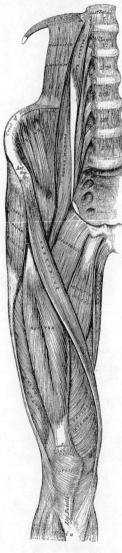

169.—Muscles of the Iliac and Anterior Femoral Regions.

dissecting dissection itself. *How would I explain this to someone,* I start to wonder, *the satisfaction that one derives from dissecting?*

The pleasure, I decide, lies in making order of the disorder, in tidying up what looks messy. It is an art well suited for fastidious types such as myself. "Make it pretty," Kim and Dana often instruct, half seriously, but it's true—that is exactly what one strives to do. When done well, dissection is very pleasing aesthetically.

With the anterior compartment completed, I move on to my final task, the dissection of the "femoral triangle" in the upper inner thigh. If you make a triangle of your forefingers and thumbs, you have approximated the size of the femoral triangle. It is bordered by two distinct muscles and a ligament (the sartorius, adductor longus, and inguinal, respectively). I remove the skin and fascia covering it, only to find a thick chunk of yellow fat. Such a sight would have made me recoil not long before. But I have gained a healthy respect for fat. In modest amounts, it serves a vital purpose, providing insulation as well as protection—padding. In fact, there is an unwritten anatomical rule (clearly written in leaner times) that if you find fat, you'll find structures in need of protecting. I put down my scalpel and, again, use my fingers, operating purely by touch. Sure enough, deep within this chunk, I feel a thick vessel. *That's got to be the femoral artery.* I do not need to glance at the lab guide; I know it. The femoral, used for coronary angioplasties, runs a nearly straight shot up to the heart. *And now I'm going to find—yes, here we go—the femoral vein and nerve.* Everything is in its proper place.

That the body is structured in such a consistent, organized way is another reason I find dissecting supremely satisfying, a view that Henry Gray surely shared. In fact, perhaps I misstated it earlier. Dissecting really has nothing to do with making things orderly. The order is already all there, just under the surface. The anatomist only has to uncover it.

HENRY GRAY IS in a box somewhere, I keep telling myself. Somewhere he survives in a box of letters, personal papers, manuscript drafts, page proofs (*something*), stashed away in a basement, a mislabeled carton, a forgotten storeroom, a locked drawer (*someplace*), just waiting to be discovered. But the box eludes me still. My many inquiries to libraries, universities, and medical societies have resulted only in the most politely worded series of *Nos.* Recently, however, two separate archivists I'd contacted added an intriguing footnote. Both mentioned another person who had made similar queries about Gray's papers. *Maybe he found something?* Unfortunately, the inquiries had been made more than a decade earlier. Fortunately archivists specialize in saving such items as old correspondence; soon, I have a name and a London address. I dash off a letter.

Just four days later, an e-mail arrives from Mr. Keith E. Nicol. Though I'd purposely kept my letter brief, simply introducing myself and expressing my sincere interest in Henry Gray, apparently this is all Mr. Nicol needed to know. "I look forward to assisting you with your research, as I have quite a lot of information on Gray and his life and career in medicine," he writes, suggesting that I begin by compiling a list of questions. "I will do my best to answer them." It's as though he had been waiting to hear from me all this time, and now he is eager to get started.

I promise to get back to him with a list of questions, but first, just one: Where had his interest in Gray originated?

He was working at a London teaching hospital in 1990, he replies, and had begun assisting a fellow staff member who was contemplating writing a biography of Henry Gray. The project didn't last long for either man. By year's end, Keith had been "made redundant" from his job, and soon after, the writer decided there was just not enough material for a book. Regardless, Keith had become hooked. Trying to piece together the anatomist's life was a puzzle he could not set aside. And though he had no writerly aspirations himself, he took over the Gray research completely. Slowly, painstakingly, he accumulated a tidy collection of facts and details, most of which he discovered through old-fashioned detective work, hunting through municipal records offices and local libraries and archives, including those of St. George's Hospital.

While nourishing his larger fascination with English history, the research also offered Keith a diversion during some very difficult times. In the midnineties, his wife was diagnosed with breast cancer and, after an eight-year illness, died in the spring of 2003. The year before Sue's death, however, he had been able to bring his decade-long pursuit of Henry Gray to a satisfying close. He had created a year-by-year breakdown of the anatomist's life, a document that documented Keith's research.

"I shall post a copy of the chronology to you tomorrow," he promises.

Till then, Keith had a single question for me, one that brings a smile to my face: "Are you aware of the link between Henry Gray and Henry Vandyke Carter?"

SIX WEEKS AFTER his mother's death and eight weeks before the anatomy book was due at the publisher's, H. V. Carter hit a wall. "I have fallen into a languid state, with occasional fits of despondency and ever-active and vagarious thoughts, 'specially of the future," he reports to his diary, May 31, 1857. As if making a diagnosis, he adds, "This state is probably somewhat morbid—the result of a single constant, solitary occupation—this drawing on paper and wood."

That he would blame the book is understandable. For sixteen

months now, he and Henry Gray had been toiling away at what, at times, must have seemed like an endless project. Simple math says that Carter had to complete two drawings every three days over the eighteen-month period, though, clearly owing to his talent, there is not a single image in the completed book that looks rushed. Tellingly, Carter writes of drawing as if it were his sole, all-consuming job, when, in fact, he had also been serving as demonstrator of anatomy throughout this period and, since the previous June, as demonstrator of histology, not to mention his tutoring for several hours each day. Calling drawing a solitary occupation, though, was no exaggeration. Save for Saturday afternoons, which he spent with Gray, Carter worked by himself at home, and, he acknowledged, the quiet sometimes got to him. I can well imagine that on Sunday, his one day off, he rejoiced in the fellowship, albeit fleeting, he felt at church.

Gray was just as busy as his collaborator, if not more. He had to produce, on average, ten pages of text every week, on top of holding three separate jobs—lecturer on anatomy, curator of the Anatomy Museum, and surgeon to the St. George's and St. James's Dispensary—as well as fulfilling his duties as a member of the Pathological Society Council, the Royal Medical and Chirurgical Society, and the Royal Society.

One glaring difference between the two Henrys, however, was financial. While Gray was making a handsome living at the hospital, H. V. Carter, by contrast, was not being paid for either of his St. George's jobs. The histology position came with no salary—he had taken it as a favor to Gray—and, shockingly, as demonstrator of anatomy, he had not been paid a pound in eight months (funding had dried up after his mentor Dr. Hewett retired). The disparity in income between them was even more pronounced in their respective contracts for the book. While Gray would be paid £150 for every thousand copies sold (an arrangement that would ultimately benefit four generations of Grays), Carter, who had negotiated his own contract with the publisher, would receive no royalties, only a onetime fee of £150. Why should he have accepted such an inferior arrangement? To him, it wasn't—£150 was three times larger than the salary for a demonstrator—but finally, his own naïveté also played a role.

As for why Carter soldiered on at St. George's, this was partly out of loyalty to Gray, I believe, and partly because he was well suited to soldiering. He also felt, at least initially, that his unpaid work might eventually pay off in his forming the right connections and, should Providence allow, a plum job. But now, he was clearly having second thoughts about staying on at St. George's. A vacancy for curator of the Pathology Museum had opened up in March—a position equivalent to Gray's at the Anatomy Museum—but, two months later, Carter still had not decided if he would even apply for it.

"I'm constantly imagining myself in another position—as surgeon in a small country village, or the like," he confides to his diary. However, faced with endless drawing, he had begun to fear that "[I'm] losing much of the little practical professional experience I had." This was a legitimate concern. Though he had recently earned his final certification to practice medicine, having passed the M.D. examinations in late November 1856, Carter had treated no patients other than his mother and done virtually no hands-on doctoring since he'd filled in for John Sawyer two and a half years earlier. Adding to his anxiety was the reality that, at age twenty-six, he was off to a late start in establishing a career.

"All my fellow students, and most since my time, have left the School and are settled in life," he observes on another date. "Whenever I hear of their advancement, or meet them, [these] active men, there is a kind of stunning sensation or pang in one's breast—and the same kind of feeling on seeing the younger students emerge into grown manly fellows." Indeed, as Carter notes, the very pupils he had been teaching would be his competitors for the same jobs and, because of their youth, perhaps have an edge. The thought is almost too depressing for him to contemplate. What never occurs to him, ironically, and never would, is that the bane of his current existence—drawing—would end up making his name. For the moment, all he can think about is bringing this "quiet mechanical occupation" to a close and putting the anatomy book behind him.

And what about the author? During the same month, as Carter wallowed in his "languid state," was Henry Gray over at 8 Wilton Street burning the midnight oil, finishing the last chapters of the book? Was the undertaking taking its toll on Gray as well?

Maybe.

Upon receiving Keith Nicol's chronology, I turn at once to the year 1857, hoping for a revelation. There is none. Alas, Keith had not discovered a cache of Gray's papers, a diary, or anything of the kind. In fact, Keith later tells me, he had found almost nothing about Gray's personality or personal life during his research. Keith did, however, track down a small collection of letters written *to* Henry Gray (now in the possession of a distant relative). In these, a curious fact emerged. At the end of May 1857, Gray took a six-month leave of absence from St. George's, during which he served as personal physician to George Granville Sutherland-Leveson-Gower, the second Duke of Sutherland, a seventy-one-year-old nobleman whose London home was a quick brougham ride away. The correspondence gives no hint as to why Gray would take this job, but to me, there's an obvious explanation. It was not for money or prestige or the chance to rub elbows with royalty—granted, all nice inducements. No, to echo the two Henrys' guiding principle, it was for practicality's sake. With the Duke of Sutherland, Gray would have just one patient (the duke lived four more years, so I suspect his demands on the doctor were not extravagant), and this arrangement would give Gray the one thing he needed most: extra time. Freed of his usual responsibilities, he could bring his manuscript, at long last, to its end.

I HAD NEVER heard a single unflattering word said of Henry Gray. Then I spoke with Charlie Ordahl.

"Over the past couple centuries, anatomists like Gray have been like the monks of the Catholic church, re-creating the same illuminated manuscript," he tells me after lab one day. Dr. Ordahl— "Charlie, please"—is one of the eight anatomy instructors. "It's the whole scribe-type tradition," he continues. "Very honorable. But that mentality is what makes anatomy as a science such a failure. Nothing changes, decade after decade. Textbooks are the same. It's taught in the same way—with cadavers, all on-the-table, same position. The same memorizing in the same order of all these parts you're never going to see again."

Little did I expect that the mere sight of my copy of *Gray's Anatomy* would provoke such a prompt mounting of a soapbox. But Charlie can stand on the book itself if he desires. I am quite enjoying his diatribe. "So, why do you think that is?" I ask, urging him on. "Why has nothing changed?"

"Well, it's partly the power of a paradigm," he resumes. "Anatomy is like gravity, to most scientists. You don't question it." In addition, he notes, after William Harvey discovered the circulatory system in the early 1600s, there wasn't really much *new* to discover in the body. It had all been found. "Anatomy became encrusted; it lacked a raison d'être."

Charlie may sound like a newcomer, one of the new breed of computer-simulation-loving anatomy instructors, but he is actually an old-timer, a white-haired bear nearing retirement age. "So if the field is rusted over," I ask, "what is the point of teaching it this way?" I look around the lab, gesture at one of the cadavers. "Is it just tradition?"

"Sure, that's a big part of it. But the *real* deal is, you can't do medicine without it. You have to have that basic understanding of anatomy. Of course, I think you can make it a *lot* more streamlined. Teach it in a much more focused way. Doctors don't remember any of this stuff. Ninety percent of what we throw at them doesn't stick."

It's a good thing Dana's not in earshot, or these two might be headed for a smackdown.

I ask how he would propose anatomy be taught.

"Well, for example, you can make a case for studying anatomy solely from the perspective of connective tissue."

"You mean like fascia?"

"Yeah. Fascia. Ligaments. Tendons. Mesentery. Connective tissue *organizes* the body. It gives you a clear organizational *structure,* especially in the limbs. If you know your deep fascia compartments—posterior, anterior, and so on—you're set. You don't need to memorize every muscle and nerve. Likewise, you can use connective tissue as your highway for looking at *all* the systems and parts.

"Plus," he goes on to say, "it makes sense from an embryological point of view to focus on connective tissue. It's one of the first things to develop in utero. It surrounds every nerve fiber. Every muscle cell. It's what holds the body together. And, of course, over a lifetime, it changes. In fact, some people say aging is connective tissue becoming tighter."

Now that's a novel way to look at getting old.

"Most of us don't give our connective tissue any thought," I point out. "It's not like bones or muscles. You're not even aware you have it."

Charlie grins. "Right. Well, you would if it were missing."

How true. Everything is connected. Every word Charlie had just said, it strikes me, could be applied more broadly. One can think about life solely in terms of different kinds of connective tissue: The attachments to family and friends that sustain you. The relationships that anchor you. The bonds that tighten with age. On some deep, unseen anatomical level, connectedness is vital. Without it, you would fall apart.

SHORTLY AFTER HENRY Gray said his temporary goodbyes to St. George's, H. V. Carter said goodbyes of his own, only his were for good. "Have now taken some decided steps which have severed my connection with the Hospital," he reports on July 27, 1857. "No longer Demonstrator. Did not apply for the curatorship, though had good interest. Indeed, the thing is done and I am not quite certain how wisely, but the monotony and uselessness of my present life, as I have myself made it, was the final inducement. And yet, I have no fixed plan for the future—one occupation gone and none other selected: Was this prudent?"

No, certainly not, he knows. But Carter felt he had no choice; the time had come to leave St. George's. This was much more than a professional crisis; it ran deep and personal. Increasingly over the past year, he had been questioning his purpose in life but was again and again coming up empty. For instance, in a remarkable passage written six months earlier, in January 1857, Carter had examined this emptiness with a dissector's eye. "My life, as I too much anatomize it, daily, or hourly, is far from happy. . . . God is hidden. Christ, I know not. . . . I am full of indolence, lonely, uncheered and unassociated, unaided, without plans or purposes and like a thing only looked at."

One of the great liberties of keeping a diary, I believe, is the freedom to indulge in self-pity without self-consciousness—to record "a thousand despondent thoughts," as Carter once put it, without worrying about anyone's keeping count. But with his January 1857 entry, a dangerous shift had occurred, in my view. He was beginning to suffer a profound loss of self. As if referring to himself as "like a thing only looked at"—an object, a specimen—wasn't a large enough tip-off, he boiled the feeling down to chilling effect in early May, a month after his mother's death: "Am daily becoming <u>anonymous</u>." No one knew him, he felt, and, worse, he no longer knew himself.

In spite of his despondency, I have never gotten the sense that suicide crossed his mind. Deep down, he was too God-fearing. What's more, he felt a responsibility to his brother and sister, as well as to those at St. George's who depended on him—students, instructors, and not least Henry Gray. But with Gray's departure at the end of May, Carter was freed somehow to make a bold move of his own. Still, he made sure, first, to fulfill his last professional obligation. And sure enough, in the same July 27 entry in which he reported his severing ties to St. George's, H. V. Carter made an equally significant announcement, speaking for himself and his collaborator: "The Book is finished." The sentence is a sigh of relief. Though he doesn't pat himself on the back, in the moment while the pen is still warm in his hand, he is clearly feeling up. He closes the entry with, "Health is good. I trust, I hope, and hope in trust."

The mood does not last. Though he is no longer a slave to the anatomy book, the feelings of isolation become even more intense.

Everything that had kept him tethered to the world is gone. Sounding panicky, he writes on September 3, 1857, "My situation is entirely critical and I as passive as ever in the tossing waves." He begins to sink. However, as Carter so eloquently puts it, "At times of depression, memory always excels." And suddenly he remembers, or so it seems, an escape plan he had once hatched: India. Dazzling, exotic India: a place where men went to make their fortunes, to reinvent themselves, to serve their country. India: a nation in the midst of a violent revolt against the *Raj* (British rule) that would come to be called the Indian Mutiny.

FIVE MONTHS LATER, Lily Carter of Scarborough, Yorkshire, receives in the post her brother's diplomas, for safekeeping, and the following note:

[London, February 1858]

Dearest Lily,

I feel there is a great change of life coming, which cannot be altogether prosperous. My dear, we must be patient and enjoy the present while we can. . . . I pray you may ever be happy.

Goodbye, dear Lily!

Your affectionate brother,
H. V. Carter

I shall write whenever possible—tell Joe I have by no means forgotten him.

PART THREE

THE ANATOMIST

*No man should marry until he has studied anatomy and
dissected at least one woman.*

—Honoré de Balzac, *The Physiology of Marriage,* 1829

Fourteen

I LEAVE FOR LONDON JUST AS H. V. CARTER LEAVES IT BEHIND. "Did feel at the [rail] station sharp pangs of regret," he admits to his diary, a printout of which I have brought along to read on the eleven-hour flight.

The train takes him to Southampton, where he books his passage to India: £95, notes. The boat, a paddle steamer named the *Sultan,* sets sail the following day, February 24, 1858. "Left at 2 P.M. Weather fine and water smooth. Troops on board. Fast officers, slow passengers, 3 ladies and 2 children, about 30 in all. Ship full." The thirty-five-day, sixty-two-hundred-mile voyage includes stops in Gibraltar, Malta, and Alexandria. They follow the path of the Nile south to Cairo and see the pyramids at Giza, then trek overland by train and caravan to Suez, where he and the other passengers board a second steamer, this one bound for Bombay, Carter's final destination. "Arrive here safe and almost well," he jots on March 29, clearly relieved to be back on terra firma.

By the time my plane lands at Heathrow, a good fifteen months have passed. It is May 15, 1859, exactly one week before Carter's twenty-eighth birthday, and I have reached the portion of his diary that, in part, brings me to London: a gap of two and a half years, a significant period in the history of *Gray's Anatomy.* But the gap in Carter's story is small compared to the void that is Henry Gray. I still know so little about the man that he is becoming less rather than more real to me. I have come to flesh out Gray's ghost.

Though this is my first time in London, the sights I most want to see are not on the standard double-decker bus tour. I therefore have come with my own set of maps—nineteenth-century and present-day—and, as important, my own personal map reader and navigator, Steve. As for my lab partner, I have left Kolja in the capable hands of Meri during my week away.

After crashing heavily, Steve and I caffeinate heavily and head straight to Hyde Park Corner. Here, the Grand Entrance of Hyde Park meets the Wellington Arch meets the western edges of Green Park and Buckingham Palace Garden. Here, tourists stop dead in their tracks and wrestle with ill-folded maps, Steve and I being no exception in this regard. Here, we find St. George's Hospital.

From the outside, the structure looks almost exactly as it does in early nineteenth-century engravings, and the hospital name remains chiseled in the cornice. But if there are any doctors inside, they are on vacation; the building is now a luxury hotel. We cross to the north side of the busy intersection to get a better angle for a picture.

Henry Gray spent sixteen years in that building, nearly half his life, I tell myself as Steve gets artsy with his new digital camera. No, make that fifteen and a half years, if you subtract the six months with the Duke of Sutherland. Gray, I recall, returned from his leave and resumed his St. George's duties on the first of December 1857, at which point the anatomy book was well into the production stage. Early

St. George's Hospital, as it appeared in Henry Gray's Day

the previous month, in fact, John Parker Jr. (the son in John W. Parker and Son) had informed the author and the artist of a problem with the engravings: some of the woodcuts were too large for the book. In Carter's account, perhaps overly self-castigating, he and Parker shared blame for the error. "His neglect is as great as my ignorance at least. Gray will clear himself." To the satisfaction of all three, fortunately, a solution of some sort was found, and production proceeded without further delay. By the time Carter left for India, Gray was likely already reading page proofs, as the book was set to be released in August, just six months away.

With the writing behind him, Gray could recommit himself to one of his long-standing responsibilities—serving on the St. George's board of governors, which, in a roundabout way, would end up having a profound impact on his romantic life. At the March 24, 1858, board meeting, he and the other governors approved the hiring of a new assistant apothecary named Hugh Wynter, and it was likely through young Hugh that Gray was introduced to Miss Elizabeth Wynter, Hugh's sister, who in time would become Henry's fiancée.

The longer I stare at the building, the more distracted I become by the cars and buses and traffic lights. But going inside the hotel ends

The same building as it appears today

up being a big mistake—like accidentally deleting an entire file of pictures on your computer. In these plush surroundings, it is no longer possible to envision any part of Henry Gray's life having taken place here.

Steve and I decide to take a walk.

Ten minutes, four blocks, and three map consultations later, we stand just outside the covered entryway to what was once No. 9 Kinnerton Street (the street name and numbering have long since been changed). The building itself is set about half a block back from the road. Though the weathered brick exterior is definitely original, we can tell that the spacious, vaulted anatomy lab on the top floor no longer exists, perhaps a victim of the Blitz or simply renovations. The building now houses posh residential flats.

We walk through the narrow passageway that students in Gray and Carter's day likened to an ear canal and head to the eardrum, or front door. At the moment, I feel less like a student and more like a proselytizer. *Do the people who live here know who once walked their floorboards?* A small part of me is tempted to press one of the dozen door buzzers and invite myself in for a chat. I would talk not only about Henry Gray and H. V. Carter but about Timothy Holmes, a charismatic St. George's man who creeps into their story. Holmes, who shared demonstrator duties in the lab during Carter's tenure, was, by all accounts, a larger-than-life character—a one-eyed surgeon, a lover of theater, a terror to anatomy students. Holmes helped his friend Henry Gray read and edit the hundreds of page proofs for the first edition of *Anatomy, Descriptive and Surgical.* But more important, Holmes served as editor of the book for seven consecutive editions following Gray's death. He, probably more than any single person over the next twenty years, helped keep Gray and Carter's legacy alive.

Rather than disturbing the Kinnertonians, Steve and I exit the ear canal and keep walking. The neighborhood is filled with fine Georgian homes, and I find that, by lifting my line of sight just above car level, it is easy to imagine we are in another time—the mid-nineteenth century, to be precise. Jet lag no doubt enhances the effect.

Checking our nineteenth-century map and following Henry Gray's likely path home, Steve and I wind our way around the crescent-shaped Upper Belgrave Street. Although I've often had trouble getting inside Gray's head, I feel certain I know how he felt while taking this walk on September 11, 1858: in a word, elated. The first review of his book had been published. Of "Mr. Gray's 'Anatomy,' " *The Lancet* editors declared, using the abbreviated title by which it would become known, "we may say with truth, that there is not a treatise in any language, in which the relations of anatomy and surgery are so clearly and fully shown." Indeed, "it is impossible to speak in any terms excepting those of the highest commendation. The descriptions are admirably clear, and the illustrations, copied from recent dissections, are perfect."

Though they had no quibble with the *Anatomy* per se, the editors, as if anticipating the book's success, felt compelled to issue a word of warning on anatomy texts in general: "No book, however ably written and accurately illustrated, can ever enable the student to dispense with the necessity of the actual dissection of the human body, and the study of disease at the bedside. . . . The student who trusts solely to books, however excellent they may be, will find himself, in the hour of trial, theoretically learned but practically inefficient." With that said, the review of the 782-page *Anatomy, Descriptive and Surgical* closed on a high note: "As a full, systematic, and advanced treatise on anatomy . . . we are not acquainted with any work in any language which can take equal rank with the one before us." And what a steal at twenty-eight shillings!

It is the kind of review a writer dreams of bringing home to Mom. Which is exactly what Henry did, I have no doubt. At thirty-one, he was the last of the Gray children still living at the family home with their widowed mother, Ann, now sixty-six. Thomas, the eldest, had been married for thirteen years and, by this point, had fathered eight of his ten children. Though barely a breath is known of Henry's two older sisters, Elizabeth and Mary, they were both presumably married by this time. Tragically, Henry's only younger sibling had died three months earlier. At age twenty-seven, Robert

Gray, a seaman aboard the merchant vessel *Indomitable,* had perished at sea on May 23, 1858.

Steve and I approach the south end of Wilton Street and take a left.

We find Henry Gray's home not by the number on the door but by a plaque embedded in the mustard-colored brick on the second story, one of the London County Council's distinctive round markers designating a historical landmark:

HENRY

~GRAY~

1827–1861

ANATOMIST

lived here

I had hoped we might be able to convince the current occupants to let us take a look inside, to creak the floorboards and see where Henry Gray had once resided, but clearly, no one lives here anymore. Through a ground floor window, Steve and I glimpse what looks like an abandoned renovation project. Most tellingly, the doorbell has been removed from its socket, leaving a hole in the doorframe. I rap on the door anyway—once, once more. *Oh, give it another try,* I tell myself. *Maybe the world's slowest carpenter is inside.* "Hallooo!" I add in a friendly voice. "Hello?"

I notice another small sign of disrepair. Where a big brass house number had once obviously been nailed to the front door, as on the other homes along the street, one is now simply scribbled, a faint penciled 8. With a few strokes of an eraser, one could wipe it away. It seems like an all too apt metaphor for Gray himself: here at home one day in June 1861, gone the next, taken, virtually overnight, by sudden illness.

Steve and I cross to the other side of the street to take in the whole façade of the building. I remind myself that Gray spent countless fruitful hours within writing the book that would bring him a kind of immortality. During the last big push to complete the work,

Carter often visited on Saturdays. He stood on that doorstep, rang the bell and greeted Mrs. Gray, I'd imagine, then joined Henry in his office, where they passed the afternoon working.

For all the time they spent together, I wonder how well they really knew each other. In particular, how well did Carter let himself be known? Compared to someone as "naturally clever" and accomplished as Henry Gray, a man who seemed to belong to a different "genus" altogether, Carter felt inferior. As he once put it, "The genus is not my natural one. I belong to another generic division of men— the one <u>below</u>." Heaven help him, H. V. Carter believed himself to be ordinary.

The sad thing is, I don't think Henry Gray saw him this way at all. In his preface to the book, Gray calls Carter "his friend," which, granted, is not terribly revealing on its own. But he did leave behind one other clue as to his true opinion of H. V. Carter, and it is hidden in plain sight. In identically sized type—no doubt in accordance with his wishes—the spine of the first edition of the book reads:

GRAY

ANATOMY

CARTER

Henry saw Henry as an equal. They were and would always be two men linked by anatomy. If Carter noticed Gray's gesture, however, he never mentions it in his diary. But then, if he saw the excellent reviews the book had garnered, not only in *The Lancet* but in the *British Medical Journal,* among others, he does not say so either. And if he were aware that sales were brisk, that the first edition of two thousand copies was well on its way to selling out, and that an American publisher had already bought the rights to the *Anatomy,* he keeps it all to himself. All of which has left me wondering, how exactly did he feel about the book?

Upon receiving his first copy of the *Anatomy* in mid-October 1858 (the book had been sent not by Henry Gray or the publisher but, oddly, by his onetime boss at the Royal College of Surgeons, dear old

Mr. Queckett), Carter's only words are: "The Book is out and looks well." *Well.* It's a muted response, to say the least. His silence makes me think he paged through it once, then put the book up on a shelf, literally and figuratively, as though it represented one long, unpleasant chapter that he just wanted to put behind him. Which could very well have been the case. In only six months, H. V. Carter's life had changed utterly.

Though he'd had to pay for his own passage to India, Carter was officially an enlisted man. He had obtained an appointment with the Indian Medical Service, the medical corps of the British-controlled Indian Army, which, at the time, provided personnel for both civilian and military posts. Immediately upon his arrival, Carter reported to Fort George, an army post outside Bombay, where he helped treat soldiers wounded in the mutiny. Before he could even learn the ropes, though, he was assigned to an artillery unit stationed in Mhow, a city in central India, 350 miles (563 kilometers) and a ten-day trip away. But no sooner had he arrived in Mhow than the military action had moved on and he was once again reassigned. This time, he was instructed to turn around and make the long journey back to Bombay, though this was complicated by troops, moving in the opposite direction, receiving transport priority. When finally allowed to leave Mhow, he ended up riding solo on a "bullock train," an oxen-driven cart. "Not much danger," he noted, "but most men on road armed," and to his astonishment, a leopard and other exotic animals boldly crossed their path. In a certain way, this was exactly what he had signed up for—an adventure worthy of Bellot—yet he could not get back to Bombay fast enough. The job awaiting him there was, as he put it with unalloyed joy, "The very thing I had wished for!" He had been appointed the anatomy professor–cum–Anatomy Museum curator at the newly established medical school for Indian students, Grant Medical College, as well as being named a staff surgeon at its affiliated hospital—positions that would instantly grant him the elevated status he craved. In other words, Henry Vandyke Carter was about to become, for all intents and purposes, a Bombay Henry Gray.

• • •

RETRACING CARTER'S EARLY steps in India is relatively easy. From the moment he disembarked in Bombay, he recorded his every movement, as though he'd placed himself under surveillance. On his trip to Mhow, he even charted the number of miles covered, day by day, village to village. Still, his words take one only so far. Typical for a diarist, he stints on atmosphere, which is a shame because, in those rare instances when he is moved to do so, Carter's diary writing can be transporting. Three days before he is to give his first anatomy lecture to his Indian students, for instance, he finds himself in a sanguine mood and takes a moment to capture the beauty of the day. "Though the monsoon [season] has begun," he writes on June 28, 1858, "the view and prospect from these quarters of Fort George towards the harbour [is] pleasant and lively. All shines outside, and the splendid home-ships ride at anchor like seated queens." In a rush of images, he describes "all those little details which serve to complete a picture": the "simple native boats . . . the passing clouds and towering hills, and variety of light and shade," and in the foreground, "the small unfinished native pier, bit of beach, and timber-logs of palms."

Somewhere between these sentences, he has an epiphany. "Truly, there is some pleasure to be found in such scenes," he writes, "and at these times when Nature shews her peaceful and smiling face. Why not then rise to the contemplation of Nature's Lord and Maker?" Indeed, why not? Though this day was Sunday, he had not attended church, and yet here, by simply taking in the view from a window, he felt the Lord God's presence. "At last His mercy and goodness are revealed to my dull vision. At last I have found Him." Completing the picture that Carter has painted is my image of him, propped against the sill, writing, happy.

Actual images of Carter are rare. Only two are known to survive, both showing him as an elderly man. So on our second day in London, Steve and I decide to play out a hunch. We visit the British Library to see what is described in the catalog as an "Album of views of 'The Grant Medical College and Jamsetjee Jeejeebhoy Hospital,

Bombay.' " The photo album, containing forty large vintage plat-
inum prints, is among the library's vast holdings of material related
to the British rule of India. While the photographer is unknown and
the dates uncertain, what makes the album sound so promising is
that it contains views not only of the buildings but also of the staff.
Could we match his name to an unidentified face?

Getting permission to see the album, however, requires an inter-
rogation of sorts, as the BL is very selective about who is (and who
is not) permitted access to its historical collections. After completing
a detailed application, you have to queue up, then go through the
equivalent of a job interview, which sounds a little nerve-racking
and, frankly, more time-consuming than we had expected. Is an old
photo album worth all this trouble?

Curiosity keeps us in the queue. We pass muster in our separate
interviews. And by midafternoon, Steve and I are being admitted to
the Oriental and India Office Collections reading room, the kind of
place for which the phrase *inner sanctum* was coined. Everything
about the room, from the tasteful furnishings and hushed atmo-

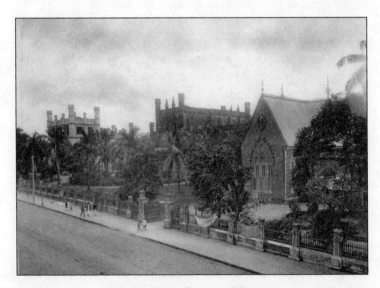

Grant Medical College, Bombay, c. 1905

sphere to the tweed-jacket-wearing patrons, screams *scholarly*—in the quietest, English-accented voice, of course.

More waiting ensues, as the requested album has to be procured, but soon we are ushered into a small, climate-controlled viewing room. We take a seat as Helen the librarian places a large archival box before us. Helen returns to her post at the door.

With the first pictures, our patience is rewarded. Here is one extraordinary, richly detailed picture after another—the college's grounds, the "J.J." hospital, the operating theater, the dissecting room, and other sites, all of which appear as dramatically different from St. George's as Bombay is from London. Even the cadavers look different, being dark-skinned rather than ghostly white. In one photo, we can practically feel the tropical heat of Bombay. These are museum-quality prints, wonderful to see and perfect in every way except one: Carter is definitely not in them. The photos were taken too late.

Regrouping back in the reading room, with only a couple of hours before closing, Steve and I request the library's one other item

Students of anatomy, Grant Medical College, c. 1905

specifically related to Grant Medical College: a series of annual reports from the latter half of the nineteenth century. I feel as if we are
scraping the bottom of the research barrel here. Having written annual reports myself, I think of them as little more than pro forma,
ghostwritten exercises in community relations. But as Steve points
out, aren't annual reports always filled with pictures?

Well, these aren't—not a single photo to speak of—though they
are filled with enough facts and financial statements to qualify them
as standard specimens of the genre. But one of them does contain
something I had never expected. In the report for academic year
1859–60, Carter himself submitted a tidy overview of the Anatomy
Department. Sounding appropriately officious, he began with the
requisite statistics (one hundred lectures had been provided, thirty-
three weekly exams, et cetera); followed with a tart review of the
twenty-two anatomy students' work in the dissecting room ("satis-
factory," though, in general, "their assiduity and zeal" lessened noticeably toward year's end); and closed with a note about how the
course had been organized. "The order of subjects in descriptive
anatomy has been generally conformed to that of the textbook,
Quain's, or [a new one] which is becoming a favourite book with students, *Gray's Anatomy,* and without vanity I may say that the figures
in the latter work are calculated to greatly assist students."

Blink and you'll miss it. I might have myself, had Steve not read
ahead and put his finger right on it—

"There's your answer," he whispers, pointing to Carter's final sentence.

And sure enough, that *is* the answer to how Carter truly felt about
Gray's Anatomy. He hadn't shelved the book; he was using it. He was
proud of it. And he was clearly pleased to see it "becoming a
favourite" with students. I am sure it pleased him, too, to see the
footnote added to his report by the school's principal: *"Dr. Carter is
the author of the beautiful plates by which* Gray's Anatomy *is illustrated."*

Along with this unexpected find comes a thorny narrative
dilemma. This is where I want to stop the story. Here is where I want

to leave H. V. Carter, with a happy ending: at age twenty-nine, having found fulfilling work and having finally found that God is always present in the beauty of the world around him. Fade out on Professor Carter in the Grant Medical College dissecting room surrounded by bright young Indian students as Steve and I exit this beautiful library.

But I can't, for I know that a very messy chain of events has already been set in motion. It started in a seemingly innocent way when Carter took four days' leave in the town of Khandalla, a trip he mentioned only in passing in an entry from mid-May 1859. After this, however, he did not write a single entry until two and a half years later, by which point he had much to get off his chest. That's all I will say for now, though. Apart from adding clarifying facts and the occasional word to assist the flow of language, I am not needed here. This is H. V. Carter's story to tell.

"LAST DATE, I noted a short trip to Khandalla," he recaps in the November 1861 entry. He then continues.

There I met a young woman passing as the wife of an officer in the 89th Regiment (Capt. Barnes Robinson, a young man of not much over 24 years)—with whom I fell into conversation. Lady-like, very lively and agreeable, though she was sick. She made a transitory impression. Returning to Bombay, while living at Fort George, I occasionally visited my namesake Surgeon Carter; there I saw her again—

"Surgeon Carter" is Henry John Carter [or, "C."], 1813–95 (no relation to H.V.), a professor of ophthalmic surgery at Grant Medical College. The senior Carter's interest in the young woman piqued H. V. Carter's.

This is when I learnt the real state of things—she was recently divorced at the Cape and not married to Capt. Robinson ["R."]. She had come over to join R. but . . . he had been suddenly sent off up country. He sent her money at long intervals and she was really in want. She had a long and severe illness—acute dysentery—and occupied rooms opposite to mine, for I had taken up quarters at Hope Hall Hotel [where the other Dr. Carter also lived] during the monsoon. Naturally, I sympathized with her condition, without communicating with her, but Carter and self were constantly talking of her. He was greatly taken up with the subject and attended to her health along with Dr. Leith and was kind to the little boy she had with her—

Her child from her first marriage.

After a narrow escape, she began to recover. As she had little money, we both wondered what she would do to get back to England. R. had almost ceased writing, probably having heard from some men of his Regiment who had passed through Bombay, visiting the Hotel, that some thing was going on between Mrs. B. and a Dr. Carter. How I became mixed up in the matter, was probably mistaken for the older man, but anyway, I began to allow self to shew some interest in her, and finding this every means was used to increase this by her, until I became more intimate than C. was.

Things came to a pass, he [C.] was not to be beguiled, . . . and I suffered myself to be so. I was then in a wretched state of mind—working hard at paper on Calculi in own room, but victim of sensuality and utter despondency. Visits were forced on me, every inducement and opportunity offered. She begged me to attend her for some fancied uterine complaint—and I yielded—even willingly. On 16th August crisis reached overpowering blandishments—and on 17th with the encouragement of a hasty kiss—she afterwards, dead of night, forced herself in my room— my God! What a night!

Perhaps Carter should have checked that all the curtains were drawn.

All this was witnessed by some railway engineers . . . who afterwards sent to me an anonymous account of all they had seen and heard, written in dogged lines, date and event.

Afterwards the course was reckless. . . . Though separate, we lived together. It became a matter of notoriety—drove out together in the buggy—everywhere. We then took fresh rooms in the Hotel, but [the new owners,] *Parsees, found things getting too hot, and in October gave her a written notice to leave (I had previously become written security for her debts and paid them all).*

. . . I found and furnished a small house nearby. . . . We lived there, not unhappily, hired carriage, bought horses, etc., etc. But conscience not asleep, often very miserable indeed. She was dissatisfied, and in December took the next step—the step even more reprehensible than before and known, so far as I am aware, to only a few individuals—made application to the Registrar (Hodge) and, swearing she was a widow, we were married at the Free Church on December 29, 1859 by James Aiken, a Scotch missionary. This was downright perjury.

Though she had inexplicably sworn she was a widow, her first husband was, in fact, very much alive. Was she now technically married to two men?

The marriage was not published in the papers. We had no visitors; the witnesses were two men hastily summoned, one a (coloured) general practitioner . . . and the other a Mr. Antone (Portuguese,) then secretary of the Bible and Trust Society. . . . I was in uniform.

Only now does he give her name.

She was called Harriet Bushell, (the name of her former husband).
Soon after I wrote to my Father and told him I was married, her name, and previous residence at the Cape, but nothing more. The circumstances which urged me to this step were my own feeling of the wrong state of things; her refusal to leave though offered a large income (£200 per annum) and constant suggestions; the risk of losing appointment and even more if things went on this way; and the hope that matters would then become straight. This, [the other Dr.] *Carter, who called once or twice, assessed would be the case. But the act was possible only because of my weakness and at the time, utter blindness of the fault of false swearing.*

Next, Carter examines his own conscience again.

> *I have a vivid recollection of driving to the Fort, joking with the Registrar, who suspecting no evil, made no opposition whatever. I believe an oath was taken and when her state had to be mentioned, the word "widow"—almost as far as I recollect improperly, though it must have been and really had been talked of between us—was written.*
>
> *Long previous to this, a letter was sent to Capt. R. partly of <u>my</u> dictation, at her request, bidding <u>him</u> adieu. . . . Also, long before this, H. had I think had an abortion. But exactly at this time she seems to have conceived, as our baby was born almost within a few days, nine months hence—this striking concurrence was not without its effect, it seemed like Providence smiling on our sin. . . .*
>
> *On 14th September, 1860 the baby (girl) was born. Only a nurse and myself present. No congratulations. It was put in the papers—no one, except Campbell of the Asylum, made any enquiry to me.*
>
> *In the winter, as money was getting short, for economy's sake (and it turned out to be no economy) we went into a house in the Bellase's Road which I furnished (£60). Sold the carriage and afterwards the shigram and horses. I had then to tramp about on foot to my work. And in January became almost sick—such a life of disquiet, indulgence, and folly.*

In late spring 1861, Harriet and H. V. Carter (who was by now thirty years old) moved for the eighth time in two years, this time into a hotel owned by Mr. W. S. Sebright Green, "a solicitor, dabbling in speculation." As rumors about Harriet could not be dismissed, Green eventually lodged a formal accusation against Carter for bringing her into his establishment.

> *At last, in July a fracas ensued. I was branded by Green as a blackguard and liar. . . . This left me terribly cut up. . . . Court of enquiry and the loss of my commission threatened.*

A "court of enquiry" is a military court that looks into military matters, such as an officer's questionable conduct. John Peet, mentioned in the next sentence, was the acting principal at Grant Medical College.

Peet then interested himself, and [the other Dr.] *Carter, who saw the real state of things. An explanation and an apology followed. But I was compelled to sign a promise that H. and I should at once live separate. Delaying, a notice to quit from the George Hotel came, and I took a house at Chinchpoogly for her, myself occupying rooms at the Hope Hall Hotel. (I had been refused entrance at the Adelphi twice.) Soon a ship was found and, the brougham and a horse being sold, £145 was paid for passage for H., children and an ayah* [a nursemaid]. *The captain of the Adripore— Hellyer by name—was briefly informed of the position of H. but I had to massage* [that is, to coax] *to get her to consent to go. Wearying sad scenes occurred, but they sailed September 20th 1861 for London.*

So where has this left him?

I still occupy rooms at the Hope Hotel—

A name that seems sadly ironic by now.

Sold the furniture and paid most of the debts. Now, have almost nothing in hand. Have written to Scarborough, only saying all had left, but giving no reason. It was understood she is to have £150 a year, no [formal] *agreement made of any kind, but by not overfair means she has possession of the Marriage certificate. Talked largely of getting a divorce someway.*

Carter writes only two more entries after this, one in January and the last in March 1862. He doesn't run out of room; in fact, he leaves more than a hundred pages blank in the diary. He simply stops. In my experience, that is how it usually goes. A diary does not come to a neat, tidy ending. The diarist just doesn't show up one day.

Fifteen

I N A WORKING-CLASS NEIGHBORHOOD IN SOUTHWEST LONDON, forty minutes by tube from the site of the original building, stands the current St. George's Hospital. There is no grand façade. No columns, no marble, no tourists. The modernist architecture of today's St. George's is straight out of the 1970s, uninspired and unmemorable. On the other hand, just as in the days of Gray and Carter, the building doubles as a general hospital and a teaching hospital and, fortunately for my purposes, contains a small archive of historical material as part of its medical library. Even better, St. George's has Nallini Thevakarrunai, the "library cataloger," as she describes herself (archivist, I would call her), who has been the soul of patience in answering my many questions via e-mail about the hospital's history.

Upon receiving her responses back home, I often thought, *What a beautiful name: Nuh-lee-nee Thu-vak-ar-roo-na;* it sounds like a musical phrase spelled phonetically. And upon meeting her in person in the St. George's library, I find she exudes a similar quality, pleasant and serene. Nallini is originally from Sri Lanka and has worked for the hospital for almost thirty years, she tells Steve and me, which presents a mystery as she does not look a day over forty. She shows us a few historic items displayed in the library (including the hide of "Blossom," the cowpox-infected cow that was the source for the first human smallpox vaccination in 1796), then leads us down a back stairwell to the archive, a small bunker of a room in the hospital basement.

Nallini mentions that she wants to let Dr. Gibson know we have arrived. Moments after she makes a call, a red-haired thunderbolt enters the room: Sandra Gibson, heir to the title once held by Henry Gray, curator of the Anatomy Museum, and a professor of biology at the medical school. After a quick volley of hellos, Dr. Gibson says, "Did Nallini tell you yet?!" Too excited to wait for an answer, she continues in her Irish lilt, "Nallini said you were coming today, so I did some looking, and I found *two specimens* that I can link to Henry Gray." The museum has only a few specimens dating from the 1850s, she adds, but by digging into old records and checking against Gray's actual postmortem reports, she had confirmed their authenticity.

It's fair to say, I am pretty shocked, knowing that the original St. George's Anatomy Museum had been destroyed during the Blitz. I didn't think any specimens had survived.

Eager to share her discovery, Sandra leads us across the hallway to "the museum," although, as she is quick to concede, *museum* is too fancy a word for the place. It is a large room filled with sturdy shelves holding hundreds of containers—bottles and vitrines—containing anatomical specimens. This is a teaching collection, Sandra explains, a resource for the med school staff, brought into classrooms when visuals are needed. We come to a stop at a low shelf in the back corner. She removes a bottle the size of a Mason jar and hands it to me. "How's your anatomy?" she asks in a friendly way.

"Pretty good," I reply, although I honestly have no idea what's inside the bottle; it looks like a dog's chew toy, microwaved.

"This is the heart of a twenty-five-year-old woman," Sandra says. "You can see the aorta here"—she points to an eye-shaped opening—"but what's unusual is, she had four rather than three aortic valve cusps."

Pickled in preservatives, the heart has shrunken over time, but I can clearly see the abnormality she described.

"Would that be what killed her?" Steve asks.

"No, in fact, Gray says in his postmortem that this had never caused her any problems or even been detected. She died of typhoid or tuberculosis or something. But it was an unusual condition, which is why he preserved it." Steve hands the heart back to Sandra, and she returns it to the shelf.

She next shows us a preserved portion of spine with two completely separated cervical vertebrae—that is, a broken neck. This is another Gray original but, unlike the heart, is still in its original container. Rather than suspended in liquid, it rests on a bed of cotton in a thick-walled glass vitrine sealed at the top with bitumen, which looks like dried black tar.

This is the closest I have ever gotten to Henry Gray himself, it strikes me as I hold the container. You could probably dust the inside of the lid for his fingerprints or open it up and hunt for one of his hairs, maybe an eyelash, and test it for his DNA. But, really, finding proof through fragments of Gray's anatomy is not necessary. As Sandra explains, both specimens can be matched to postmortem reports written in Gray's hand.

She heads out the door, and we follow her *follow me*s. Back in the archive, Nallini rejoins us, and the four of us stand before a wall lined with large leather-bound books. The topmost shelf holds what looks like a set of encyclopedias for giant children but is in fact a series of nineteenth-century postmortem reports, bound and arranged by year. Sandra had previously pulled the hefty 1858 volume, which now rests on a library cart. Nallini moves it to a nearby table, Sandra turns to a report about twenty pages in, then each takes a small step back. The two women—one pale and freckled, the other olive-skinned— wear matching expressions: *Well, go ahead, take a look.*

The first thing we notice is his signature, *Henry Gray*, underlined twice at the bottom of the page. I instantly compare it in my mind to H. V. Carter's, whose signature is both less legible and fussier-looking; Gray's penmanship, by contrast, is easy to read. The patient had been "under the care of Dr. Page" (a familiar name from Carter's diary), and in describing the condition of her heart, Gray had written, "The aortic valve was composed of four flaps." In the margin, he had noted in a smaller hand, "Specimen showing the Aortic valve is preserved in the Museum."

Sandra allows us a moment to marvel, then tells us to flip to case number 199. Here we find a report for one "William Parry," who, as Gray reported, had "fallen, head first, a height of about 14 feet—"

"Ouch," I think aloud, "that would've hurt."

"Not for long, though," Steve adds. As Gray noted, Mr. Parry "had lost all power of motion or sensation in all the extremities and in the trunk of the body." He was paralyzed and died two days after being admitted to the hospital.

Sandra has to dash off to teach a class, but Nallini invites us to pull whatever volumes we want from the wall. There are hundreds of reports by Henry Gray within these books, she tells us, and points out the two worktables on the opposite side of the room.

My first impulse is one I almost feel I should suppress: to see the postmortem report on Henry Gray himself. Without knowing why exactly, this seems ghoulish; it's one thing to read reports on total strangers, but on someone you've come to know? In any event, I am spared any further moral ambivalence. Steve is already up the step stool retrieving the 1861 volume. We find no report for Henry Gray, which, upon reflection, makes sense. Merely by looking at the thirty-four-year-old man's ravaged body, his death by smallpox would have been unmistakable, and, given the risk of contagion, an autopsy probably would not have been allowed.

Suddenly, the horror of what Gray went through hits me. He must have known, from the moment he saw a patch of pustules on his body becoming confluent—meaning, running together, a continuous blistering—he would not survive. Once the pustules spread into his mouth and throat, slowly suffocating him, the end would be terrifying. Whereas his death had once seemed incredibly fast to me, coming just a week after falling ill, now it did not seem fast enough.

WE KNOW THAT he was buried in a private grave, dug to an extra-deep depth of eight feet (not due to smallpox, interestingly, but to allow for additional interments on top); that the exact time of burial was half past one o'clock on Saturday, June 15, 1861; and that the total burial cost was £7.3s. Keith Nicol has all the documentation from the London Cemetery Company. We also know that a woman named Ellen Connor—a private nurse, in all likelihood—was present at the time of death. This fact is listed on Henry Gray's death certificate, a copy of which was easy enough for Keith to track down at the General Register Office. We know, too, that Gray had apparently contracted smallpox from the nephew he had been treating. Still, Steve, Keith, and I find ourselves puzzled by the most basic question about Gray's death. As the death certificate clearly states, he had been "vaccinated in childhood" against smallpox, so why did he become ill?

"I think the reason Henry fell prey to it," Keith suggests, "is that he'd been nursing Charles for so long and he was literally exhausted." Gray's work ethic may have also played a role. "Because he was such a hard worker as well, perhaps his general strength wasn't a hundred percent." Plus, "We just don't know how good or effective the smallpox vaccine was back then." Perhaps he should have been vaccinated again, as an adult?

Another possibility, I offer, is that the strain was especially virulent. "But if so, how did his little nephew survive?"

"Maybe he didn't get it from the nephew," Steve counters.

Keith shrugs and smiles a sympathetic smile that says, *I know, I know. I've been dealing with questions like this for fifteen years.*

As we speak, we are surrounded by the product of Keith's labors: dozens of binders containing the research he had gathered for his Henry Gray chronology. Some are still lined up on bookshelves here in his home office, but most lie in piles around us. Our visit to Keith's South London home was meant to be purely social, a chance to get acquainted over a cup of tea. But other than learning that he is an uncommonly gracious man, that he has two grown daughters, and that he is tall, bearded, and "of 1956 vintage," I do not know a lot more about Keith personally than I did before we arrived. Instead, we have spent the last three hours going through the binders and hashing out the two Henrys. This no doubt reveals a lot about the two of us. Keith was first drawn to the history of Gray, and I, to the mystery. Our research paths have taken us in different directions, mine straight to Carter, his circling lesser known figures at St. George's. But in the end, it is the story of Gray that brings us together. Here we are, in the dying light of a humid October afternoon, trying to re-create the anatomist's final days.

I tell Keith that Steve and I had spent the morning at St. George's and mention the likelihood that an autopsy had not been performed—

It wasn't just Henry Gray that was potentially infectious, Keith points out. "With smallpox, the blisters would have a very watery discharge inside them, and the skin would be stretched to such an extent that the slightest touch would break it. It would then burst as an aerosol into the air. Eventually, once the smallpox was confluent, he was horrible. So was the room that Gray was in; it was full of these fomites, which was the aerosol infection. So the bedding, the wallpaper, the curtains, everything that was in the room, was potentially infectious. Imagine that this was Henry's bedroom, the room we're sitting in—"

Steve and I look around the packed, closed room.

"The Victorians had one solution for that level of contamination: fire."

"*Fire,*" I repeat to myself. The word crackles.

"In Victorian London, there was an official called the inspector of nuisances—"

"*Nuisance* must have had a different meaning then," Steve interjects.

"Oh yes, it wasn't the guy that pesters you, saying 'Do you want to buy any naughty postcards?' " He laughs. "No, *nuisances* as in epidemics, infectious epidemics. Smallpox, chicken pox, all that sort of thing. This group was responsible for disinfecting or cleaning out the infection from a house or an area. They would probably have gone in and just said, 'Right, strip this room down to the bare plaster and just burn everything.' "

Keith hesitates for a moment. "This is just a theory, but I reckon that's what carried away the evidence that you and I so desperately want—"

"His papers." I can see them going into the flames. "His letters, diaries—"

"Possibly, yeah," Keith says, hedging a bit. "I don't know for sure."

"No, I'm sure you're right," I respond. "That's what happened to his new book, the one on tumors—"

"And his revisions for the next edition of *Gray's*," says Steve, tossing more fuel on the fire, "his original manuscript."

His clothes, the rugs, his Bible—we heap everything on the pile.

"Yes," Keith nods. "A bonfire of everything he had touched."

Sixteen

THE WELLCOME LIBRARY DOESN'T HAVE A SINGLE MS. WHEAT. NO, it has *four* Ms. Wheats, one of whom is a silver-haired man. The staff person presiding over the Special Collections Room has changed throughout the day: the petite brown-haired young woman who greeted us at 9:30 and gave us the first folder of letters from the Carter archive morphed into a middle-aged man, who, next time I glanced over, had turned into a fortyish matronly type—a cousin to Valerie Wheat—and then, wordlessly, back into another young woman, this one with ruddy skin and a nose piercing. But—who knows?—maybe there was yet another librarian sitting there in between those last two. Since arriving, Steve and I have hardly looked up from the piles of letters.

Every page of every letter—every scrap of paper in the Carter Papers, including actual scraps of paper—bears a tiny penciled number identifying its proper place in each of eighteen separate folders. Despite the impeccable organization, however, there's no math that makes it easy to calculate how long it will take to get through a single folder. It depends as much on the number of pieces as on their readability, length, and relevance. There are more than three hundred items in the Carter archive, and over the next few days, we have many loose ends to tie off. Our unexpected discovery at the British Library, for instance—that Carter had not, in fact, put *Gray's Anatomy* behind him in coming to India—has made me rethink another assumption: that he'd put his old friend Henry Gray behind him as well. Now I have a new thought. Gray *must* come up in Carter's correspondence, but where and when?

Near the bottom of a box of letters to Lily is the answer: The eighty-eighth of 116 letters, dated October 10, 1861, to be precise: "You will know ('young') Mr. Gray is dead," he tells his sister, adding on a sorrowful note, just as he was "on the threshold of a high career." As his phrasing indicates, he is not breaking the news to Lily; she would have heard or read about it long before word reached him in Bombay. Rather, H.V. is tacitly sharing his grief, which I find all the more moving. Lily had visited her brother in London twice during his years there and perhaps had met "Mr. Gray" in person. Better than anyone, she would know how keenly he felt the loss of this extraordinary man.

As an object, the letter itself captures the delicate emotions conveyed. The opaque onionskin stationery is as thin as tissue paper, and the ink, once brown perhaps, has faded to a gold so faint as to be nearly invisible. *Read it quick and write it down fast,* I tell myself, *before it disappears.*

Gray's passing is not the only sad news he shares. He tells Lily that Mr. Queckett at the College of Surgeons has also died. Carter then carefully, very carefully, drops a bombshell. "Misfortunes," he writes, "have at length broken up my little household. She who was my wife has left India, in a sailing vessel, for England, and I am now quite alone."

He doesn't give Harriet's name, as if it were too distasteful to include in so gentle a letter. As portrayed by Carter, she is continually the villain in the tale—a liar, a loose woman, a corrupter—someone I had pictured as being irresistibly sensual and impossible to please. Which is why "hearing" her voice elsewhere in the Carter Papers, starting with two letters from Harriet to H.V., comes as a surprise. On paper, she seems *lady-like* and *agreeable*, the very words Carter had used when he first met Harriet back in Khandalla.

"My dear Henry," she writes in a short note sent the day before setting sail for London:

> I am much obliged to you for giving me the Certificate of our Marriage, and I promise never to show it to any one, or to name such a document

*as being in my possession unless I am actually obliged to do so for self
protection.*

Yours,
Love,
H. Carter, Late: Bushell

Harriet writes again two days later, September 21, 1861. Now on
board the boat with her two children, she bares her soul. "I have
given you much pain and trouble, forgive me, I pray you. . . . I am
sincerely sorry that I have managed so badly."

Later in this letter, she unwittingly reveals an unexpected side to
H.V. and to their relationship. "Your last words that you would 'be
with me in spirit' are indeed a consolation. Never for a moment have
[sic] any thing taken my thoughts from you, my preserver." Harriet
then speaks of what a comfort their daughter is to her. "I can feel
that you are actually present with me in her."

So what was the real story here? Someone wasn't being entirely
honest, whether H.V. or Harriet Carter, or perhaps both. How to uncover the truth?

Steve and I start where all the trouble began, with one word:
widow. Sure enough, there it is on the couple's marriage certificate,
which we found tucked in a small file of Carter's miscellaneous papers. Why had Harriet lied? What had been going through her mind?

Her lawyer, in a letter from May 1862, offers an explanation: Harriet had had "doubts" as to whether her divorce had been finalized,
so, in order to "avoid debate," she had sworn widowhood instead—
simple as that. But why remarry if you are the least bit unsure you're
actually divorced? Something fishy was going on, and Steve and I followed the trail back to the tip of Africa, the Cape of Good Hope, five
thousand miles from Bombay. Here is where Harriet's first husband
divorced her on grounds of adultery, it turns out, presumably with
the aforementioned Captain Robinson. Now, whether she had actually been unfaithful is unknown. But *shame,* not doubt, must have led

her to lie on the marriage certificate. Far better to be known at the time as a widow than as an adulterous divorcée.

Just as this was beginning to sound like musty Victorian melodrama, out of the blue came a voice of reason: "I have already informed you that the first marriage was completely annulled and that the second marriage was valid—"

Actually, this was Steve, whispering into my ear as I typed on the laptop.

He had dug up a letter from Carter's lawyer, Richard Spooner of Kent, England, who clarified, once and for all, the legal matters in the case of *Carter v. Carter*. His verdict? H.V. had no case against Harriet. At Carter's request, Spooner had also asked Harriet personally if she would agree to a divorce.

> *Without any hesitation she said that she never was really attached to any person except you—that she loved you dearly—would do whatever you wished and never spend a Rupee without your permission—and that she would prefer living with you although she might be coldly received or received not at all by the Bombay community.*

Spooner then adds his own two cents:

> *I would venture to suggest to your consideration whether it would not be advisable to treat her kindly and let her live with you again. After a few years all former occurrences will be gradually buried in oblivion.*

DAY TWO AT the library dawns. Steve and I are joined by fellow hunters and gatherers at the large center table in the Poynter Reading Room. Although we exchange little more than nods of hello, I feel we could not be in more genial, civilized company. Directly across from us sits a middle-aged woman, both chic and bookish, who was also here all day yesterday. A Mona Lisa smile never left her face as she endlessly inputted from a stack of small medieval-looking tomes. Steve and I have decided—based on no facts whatsoever—

that she is a historical romance novelist tracking down period details and atmosphere. The young married couple is back as well. They seem as passionate about each other as they do about tracing their genealogy. As for Steve and me, we still have our own little family history to sort out: Did H. V. Carter follow his lawyer's advice (and Harriet's wishes) and take his wife back? Or did he divorce her? And what happened to his daughter? In the second-to-last entry of his diary, he writes with great anguish about the child, "that poor infant, a sweet healthy babe."

> Can I ever expect—hope—to see her again? Years ago, poor old grand-father Barlow, when he used to set me part of the way home from Dry-pool, would peer anxiously those dark Saturday nights under every passing woman's bonnet—a scrutinizing gaze so peculiar and keen that, unobservant youth as I was, it almost appalled me. He expected—hoped?—to see the face of his own daughter—a creature of the town, notoriously public—

Illegitimate.

> Am I ever to endure <u>his</u> . . . bitterness? But this is far anticipation—God knows alone our future rescue or doom. Every passing child wrings my heart.

Among Carter's papers, I come upon a small note that must have just about broken it. Dated 1867, it is written in a child's scrawl: "Dear Papa come soon."

His daughter would have been six years old at the time.

Did his heart begin to soften, finally? Did he ever find a way to forgive Harriet? The answer would be unknowable, it's safe to say, were it not for one person—

"My Dear Sister—"

Yes, Lily, confidante to all, quiet keeper of all things. This batch of letters was not from H.V., however, but her sister-in-law, Harriet. Apparently, the two women had become close after Harriet returned on her own to England.

"I wished to tell you that Henry is now the Principal of a hospital in Bombay," Harriet writes to Lily in December 1879. You can hear the pride in her voice.

Unfortunately, Lily's side of the conversation is silent. Though 80 percent of the material in the Carter Papers comes from her, not a single letter in her own hand survives. Still, it is clear that Lily and Harriet were on very good terms.

"Henry is so much engrossed in his book that we see little of him," Harriet confides in a letter from January 1881. "I for my part shall be very thankful when the said book is in the press."

Only three of these sister-to-sister letters survive, none earlier than 1879, but in this small cluster comes a remarkable number of answers. First and foremost, Harriet and H.V. were never divorced, but they also never lived together again. Instead, the couple maintained an unconventional relationship that endured for more than twenty years. Other than a series of furloughs, H.V. remained in India, while Harriet and the children lived in Europe, including England, Germany, and Italy. Mother, father, and daughter apparently did reunite on occasion, spending a month together in Rome, for instance. And at least once in India, husband and wife spent time alone together, though in a town a good distance from Bombay and its late-Victorian mores. Was this a romantic rendezvous? As it should be, perhaps, the level of intimacy between the two is never defined.

Harriet, it turns out, was not the only bearer of news about H. V. Carter and family. Lily also received delightfully chatty letters from her niece, Harriet and H.V.'s daughter, Eliza Harriet "Lily" Carter. In one of the four surviving letters, we hear for the first time about her half brother, Harriet's heretofore unnamed son, John, who was a couple of years older. Around age twenty, John ran off to Australia, perhaps seeking his fortune in the gold trade. "Let us hope that no news is good news, and that he is doing well in Australia, for we think he is still there," Eliza tells her aunt. "We often hear of boys who behaved in a similar way and yet came to no harm." John's ultimate fate is unknown.

A charming instance of her father's acting fatherly comes in a letter dated October 3, 1878. Writing from Switzerland, Eliza and her mother are en route to Florence, where she would spend the winter studying Italian and taking painting and voice lessons, she tells her "Auntie" Lily, "to try and satisfy dear Papa's wishes in occupying my time in 'the pursuit of knowledge.' " Her groaning lifts right off the page. From the sound of it, H.V., in spite of his absence, was trying to instill in Eliza his own love of learning, and she was reacting with all the enthusiasm of a typical teenager.

By this point, Carter's work for the Indian Medical Service had taken him out of the classroom entirely, and in his midforties, he had found his true calling in life: independent medical research. This seems such a natural fit for Carter, but only in hindsight. When he had received his first copy of the *Anatomy*, he confided to his diary that he could never take on such a huge project again, unless working under a "ruling mind" such as Henry Gray—a natural leader, someone able to see the big picture, to use a modern phrase. Of himself, Carter wrote, "[I] analyze life on too small a scale." What he meant as a putdown, however, I see as his great gift. Carter's ability to focus on the small, to break things down, to mentally dissect—the same ability that made him so miserable on personal inspection—is what made him such a precise anatomical artist and such a natural researcher. This man who so firmly believed "I can't" is now recognized by medical historians as a pioneer, the first scientist to apply modern methods of scientific research to the investigation of tropical diseases.

CARTER'S FIRST SIGNIFICANT finding as a researcher dates back to 1860, when he was still teaching at Grant Medical College. His clinical observation of a condition known at the time as "Madura foot" had deeply troubled him. The disease seemed to afflict only poor Indian laborers, who, for reasons unknown, developed enormously painful and disabling masses in their feet and/or hands. No treatment was effective, short of amputation. Puzzled, Carter began examining surgical specimens with the microscope he had brought with him from London, and he became convinced that the culprit was a fungus of some kind. Since the laborers worked in their bare feet or with bare hands, the organism, Carter theorized, must be entering through cuts in the skin. Though unable to prove this by growing cultures of the fungus, he published his research, and two decades later, his theory was confirmed. Madura foot eventually came to be known by a new name, "Carter's mycetoma."

After five years at the college, Carter was relocated one hundred miles south of Bombay to the district of Satara and was appointed the "civil surgeon" (chief medical officer) and superintendent to the "gaol" (jail). He spent nine years here. Not one letter written by Carter survives from this entire period, giving the spooky impression that he had been locked away himself, but the reality is, he kept himself as busy as possible. On top of his other duties, for instance, he volunteered for what sounds like a daunting task: analyzing data that had been collected on eighty-two hundred Indian lepers but was simply gathering dust in government files. The result of his work, published in an 1871 Bombay medical journal, helped to dispel a number of misconceptions about the dread disease, whose true cause (a mycobacterium) was still unknown. No, he concluded, external factors such as local geography and topography were not relevant to etiology, and even though the afflicted were almost always the poorest of the poor, neither was poverty. As historian Shubhada Pandya recently noted, "Carter took pains to point out that want and deprivation were *consequences* rather than *precedents* of leprosy." Carter also soundly refuted the imperialist notion that poor hygiene among the "natives" was to blame. On the contrary, he noted without condescension, the Indian people bathed once a day—*just like you and me,* he seemed to imply—and "personal cleanliness is not neglected."

If one were looking just at H. V. Carter's curriculum vitae, his interest in leprosy would seem to have sprung directly from that mound of raw data. Perhaps he had also encountered cases in Satara, one might surmise. But, in fact, a casual mention in his diary a decade earlier shows that Carter had begun looking into leprosy long before. From a public health perspective, there were certainly compelling reasons for him to study leprosy, but I am sure it resonated in a personal way as well. Leprosy is invoked many times in the Bible—the prophet Elisha bathing a leper, the parable of Lazarus, Jesus and his disciples healing the afflicted—and the chance to make an impact on the lives of lepers would have struck him as a worthy, Christlike endeavor.

After completing his term in Satara, Carter was granted a three-year furlough, which he used to conduct a fact-finding mission on how cases of leprosy were managed in other countries. He traveled to western Turkey and southern Europe but spent much of his time in Norway. During this remarkably productive period, 1872–75, Carter also wrote two influential books, both of which he illustrated: a monograph on mycetoma (1874) and a volume on leprosy (1874), one of the first scientific treatises on the disease. What makes this latter book historically significant is Carter's advocacy for the findings of a fellow scientist, Gerhard Hansen, whom he had met during his furlough. Hansen, a Norwegian physician, had accumulated convincing evidence that leprosy was caused by an acquired bacterial infection and was not, therefore, a hereditary disease, as was widely believed and reported at the time. Carter's book included translations of two of Hansen's latest research papers (the first time they would appear in English), which helped disseminate the findings to a broad scientific audience.[1]

Six years later, H. V. Carter had the stage to himself. The setting was London, August 1881. At age fifty, he stood before an audience composed of the leading scientists of the day, delegates to the seventh International Congress of Medicine, Louis Pasteur, Robert Koch, and Joseph Lister, among them. These men were his peers. (And they were exclusively men. Queen Victoria, fiercely opposed to equal rights for women, had threatened to withdraw her royal pa-

---

1. Interestingly, while Hansen's discovery led to effective treatments for the illness, science still cannot explain the specific mode of transmission. Leprosy is now called Hansen's disease.

tronage if any "medical women" were admitted to the congress.) Carter, now the principal of Grant Medical College, head physician of the Jamsetjee Jeejeebhoy Hospital, and a surgeon-major in the Indian Medical Service, had earned a spot in "the genus" above, to borrow his earlier phrasing, whether or not he saw himself thusly. He had been invited here to give an address on his recent discovery of the organism that caused "relapsing fever," an often deadly illness that had swept through Bombay during the Great Indian Famine of 1877, and on his related findings on another blood-borne bacterium. This body of research, which would become the subject of Carter's next book, a treatise of nearly five hundred pages, had already cemented his reputation as one of the world's leading experts in epidemic disease. By now, the fact that H. V. Carter was the original illustrator of *Gray's Anatomy*, currently in its ninth edition in England, had become just a footnote in his career.

Despite the prestigious setting, Carter's moment in the spotlight must have been rather bittersweet. In his personal life, he was now very much alone. Harriet had recently died (apparently suddenly, though the cause and date of her death are unknown), and during his twenty-three years away from England, Carter had lost not only the few close friends he'd once had but family members, too. Both his father and brother had passed away—Joe at just thirty-five and only three months after marrying his longtime love. And now, his own health was poor. While investigating the nature of relapsing fever, he'd "had the benefit," as Carter once wryly put it, "of repeated personal experience of this fever." What's more, he had contracted pulmonary tuber-

H. V. Carter, artist and date unknown

culosis. But the trip would at least end on a high note. After spending several weeks in London, he would take the train to Scarborough to see his dear sister, now Mrs. William Moon and the mother of a small brood.

Carter did subsequently return to Bombay and resume his responsibilities with the college and hospital, but following his retirement at age fifty-seven in July 1888, he came back to Scarborough for good. Brigadier-Surgeon H. V. Carter, M.D., was granted the honorary rank of deputy surgeon general for his "eminent service to medical science" and appointed honorary surgeon to the queen. He bought a large house within a stone's throw of Lily's, and within a few years, he had a new family of his own to fill its many rooms. In December 1890, at age fifty-nine, he married Mary Ellen Robison, twenty-five years his junior. The couple had two children, a son named Henry Robison (born 1891) and a daughter, Mary Margaret (1895). But Carter's late-blooming happiness was cut short. He had never fully recovered from tuberculosis and died at home on May 4, 1897, just shy of his sixty-sixth birthday.

OUR TIME IS almost up. We have been here four days in a row and are going home tomorrow. The library is closing in twenty minutes. The romance novelist has gone, leaving just Steve and me at the table. Still, we speak in whispers, purely out of respect for the reading room itself.

Sue, our favorite Ms. Wheat—a Mary Tyler Moore with an English accent—delivers two last archival boxes, one for each of us. We have already read everything—every page of every extant item, down to Henry Vandyke Carter's last will and testament—but we cannot leave London without seeing H. V. Carter's actual diary.

The second volume is slightly larger than the first, as if he had splurged on the extra half inch, but, my word, they are both so small. How did he fit so much life onto such tiny pages?

"Thank goodness we saw this on microfilm first," I say to Steve, "or I don't think I'd have gotten much past the first page."

Carter's handwriting, an endless series of trembling lines, looks more like an EEG reading, one that only we know how to interpret. There is his opening note on the fate of his first diary, how he'd had to destroy some of those early pages, followed on the next page by his peppy epigraph: "Let the same thing, or the same duty, return at the same time everyday, it will soon become pleasant," which has always struck me as absurd.

Both volumes have been rebound, replacing what must have been tough original covers, for the pages show little wear and tear. Carter

himself was not just neat but freakishly neat. There are no stains, scarcely a smear, and not a single dog-eared page. None of the ink has ever run, as if the man never shed a tear.

Steve and I trade volumes. I turn to Carter's last diary entry, written on January 9, 1862, such a bleak, unhappy time. "The unfolding of the future—the immediate future—remains, and the present hardly improves," it begins. The entry is a page long, the tone desultory. You can tell his heart's not in it anymore. "Working incessantly—the only relief," he writes in his closing paragraph.

It's a wonder that the diary has survived, I cannot help but feel, that the volumes weren't destroyed, whether purposely or inadvertently, or lost; a small miracle, perhaps even Providential, that they ended up in my hands.

I carefully thumb through the diary one last time. As the thousands of words fly by, it is not a line of H.V.'s but instead one of Joe's that comes to mind:

What is past—the past—does not, nor <u>will</u> it, detach itself and remain where it was (or where it might have been <u>intended</u> to have remained) but it must bring itself forward, and smilingly, or otherwise, present itself as an old friend.

Seventeen

L AST NIGHT I HAD ONE OF MY REOCCURRING DREAMS. NO, NOT the one where I'm doing house chores with celebrities, but the dream of flunking my high school geometry exam. It's a classic anxiety dream, in which I have completely forgotten that the final exam is today—right *now*, in fact—and I have not studied at all. In a variation on the theme, however, this one was set not at Gonzaga Prep but instead in UCSF's Cole Hall. Upon waking, I felt panic and relief—such a sour mix of emotions—then bemusement. What I'd had was a *sympathetic* anxiety dream, I realized. Tomorrow is the med school anatomy final, you see, and today, I am helping Meri study for it.

We meet at 3:00 P.M. and find the lab filled with fellow cramming students. Dana, Kim, Dhillon, Charlie, and the other teachers move among them, tutoring small groups. Our table, number 24, has been taken over by a group of guys studying our cadaver's shoulder joint, so Meri and I find a spare prosection of a leg to work with. This is one sorry-looking specimen. Were it not already dead, I would say it looked studied to death. *Well-utilized* is perhaps the better phrase. I can't help wondering how many students over the years have performed the anatomical equivalent of musical scales on this severed leg, calling off the sequence of muscles, nerves, arteries, and veins.

Meri, continuing the tradition, gamely takes the long limb into her hands. "Rectus femoris," she states confidently, plucking the prominent muscle running straight down the front of the thigh.

"Indeed," I confirm.

"Okay, and this"—Meri fingers an inner thigh muscle—"vastus medius?"

"Medialis," I correct.

"Right, right, right. Vastus medialis. So this one"—she moves to the outer side of the leg—"is vastus lateralis. And then, down here, underneath rectus femoris, is"—she points to a slender slab of muscle—"vastus intermedius."

"Exactly. You've got it."

From here, I steer Meri through the arteries and veins, but honestly, by the time we get to the nerves of the lower leg, we have switched roles, with Meri teaching me the material I missed while in London. And as we move on to prosections of the arm and hand, it is clear that she knows far more about these parts than I do. Through her other classes, Meri, like all the other med students, has learned about myotomes and dermatomes and the actions of each muscle group and the clinical implications of injuries to various parts. In the lab, I now have little more to do than hold the prosected hand and, in a figurative sense, Meri's hand as well, offering her exam-eve encouragement. Really, though, there is no reason for her to be nervous. She's going to ace the final, I can tell.

For Meri, Kolja, Marissa, and the others, the conclusion of this course marks a mere first step in their studies, a foundation they are already rapidly building upon, but for me, this is an epilogue of sorts. I am not interested in studying more anatomy, going further in the field. Instead, I would like a better understanding of how the human body came to *be*, how it became what it *is*, this complicated, magnificently designed structure. I would like to study evolution.

Knowing I am not really helping Meri much, I suggest that she join one of the tutored groups, and at a nearby table, we find Kim quizzing a clutch of students gathered around a brutally dissected body. I watch for a while as Kim, gently, wryly, authoritatively, puts them through their paces: "David, which nerve innervates this muscle . . . here?" "And, Meri, what spinal nerve segment would you say it originates from?"

I slip away to take a last walk around the lab. In one of the back corners, I meet a fourth-year med student, Barry, whom I've never seen here before. Barry, the kind of heavyset, apple-cheeked guy you might call roly-poly, explains that he is doing a monthlong elective in dissection as preparation for his upcoming surgery internship. Splayed out before Barry, who's seated on a low stool, is the upper half of a cadaver, transected at the waistline. He has slit open the abdominal cavity, revealing half the stomach, the small intestines, and half the liver. At this moment, he is examining the ducts of the gallbladder. "Getting some practice," Barry tells me. "I'm planning on being a general surgeon, and taking out gallbladders is your bread and butter."

Something about his food metaphor doesn't quite sit well with me, given the gruesome sight laid out before him, but I know what he means. "Of course," Barry adds, "in the OR, it won't be anything like this. With laparoscopic surgery—which is how gallbladder surgery and a lot of procedures are now performed—you don't need to cut open the whole abdomen. You just make tiny incisions into the stomach, thread in the cameras, snip out the organ, and you're done." Barry glances at the disemboweled cadaver with a rueful look. "Hardly ever see things like this nowadays. In fact," he points out, "transplant and heart surgeries are some of the only times you'll ever open up a body like this."

I ask if he remembers learning a lot in his first-year gross anatomy course.

"Well, the thing is, you don't really *learn* it till you have to *use* it. Before that, you're just memorizing great bodies of information without being able to apply it—"

"But *that's* not all that studying anatomy is about," I counter, a bit more emphatically than I had intended.

"Yeah, true, there's a 'rite of passage' to it—going through someone's *body* with your hands. Your own two hands. Almost a ceremonial aspect to it." Hundreds of years ago, he adds, they didn't even wear gloves. "Students had to dissect bare-handed."

"Well, I'm glad I missed that era," I admit. As for Barry, he doesn't look quite so sure.

I cross to the other end of the lab, where Matt, a classmate who'd worked at table number 22, is studying by himself with a lower limb prosection. I ask Matt how the class had gone for him.

"Fine, I think I've done all right," he answers reflexively, then pauses and gives it more thought. "It's amazing you can get into medical school and *not* know what side of the body the liver's on," he says—by "you," clearly meaning himself. "Or how big the lungs are—"

"But now you do," I say.

"Yep." Matt, blond and blue-eyed, the epitome of a midwestern all-American boy, shakes his head sheepishly. "And now I know, if some-one has a pain on this side"—he jabs at the left of his stomach—"well, it sure isn't appendicitis."

I ask him what kind of medicine he's planning to study.

"Pediatrics, probably." He glances down at the prosection. "Definitely not surgery."

I wish him luck and go over to the sink to wash up. All around me, from every corner of the room, I hear the sound of *teaching,* the clear, impassioned voices of the instructors and TAs: "The radial artery goes through the snuff box and gives rise to the . . ." "Also, your thenar eminence performs this motion. . . ." "There are eighteen intrinsic muscles of the hand, but they're in groups, so they're easy to . . ." "What's the mnemonic for the rotator cuff? Right, SITS. . . ."

And rising above the clatter of voices, I hear the distinctive sound of my first anatomy instructor, Dana, the passionate anti-mnemonicist: " 'Why?' Well, first of all, you can't ask why," she is saying in a high-pitched stampede. "It just *is.* That's how we're made, but . . ."

Smiling, I head for the back corner to retrieve my bag. Just as I am gathering up my stuff to leave, a group of students converges around table number 24. "Okay, now the three muscles that attach here at the pes anserinus?" the TA asks them, pointing to the knee dissection I had performed.

"Sartorius, gracilis, and biceps femoris," I whisper to myself.

"That's right," I hear him say as I walk toward the door. "And the mnemonic is SGBF: Say Grace Before Food."

Amen, I think, as the door to the anatomy lab closes behind me.

Epilogue

TWO YEARS PASSED BEFORE I RETURNED. I WAS NERVOUS BEFORE-hand, nervous in a way that brought to mind the very first time I had made the trip. I could easily remember stepping off the elevator, rounding the corner, and heading down that dim, narrow hall, look-ing for room 1320. The hallway had seemed to get dimmer and nar-rower the farther I got, if only because there was a bottleneck of students at the door—pharmacy students—the half of the class who'd accepted Sexton's invitation to visit the anatomy lab a day early. It wasn't that the door was locked. The problem was, no one wanted to go first, the unspoken fear being that an initiation was about to begin. Inside that lab over the next ten weeks, you would be forced to confront your innermost anxieties about death and dying while taking apart a dead body—an emotional vivisection, of sorts. How would you handle it? I wasn't sure how I'd do myself.

I remembered hearing a young guy behind me ask another, "Have you ever *seen* a dead body?"

"Um, yeah, but I've never *touched* one," guy two answered, sound-ing none too thrilled at the prospect.

Henry Gray and H. V. Carter would have gotten a kick out of that. As young men, those two certainly did not enter the Kinnerton Street lab expecting to learn life-changing lessons about mortality. They did not need them. Gray and Carter had each seen and touched plenty of dead bodies before they began dissecting them. And when they did dissect, many of the cadavers were likely close to their own ages. In the early nineteenth century, in England as in the United States, the life expectancy at birth for a male was half as long as today—just thirty-eight years. For a female, it was only two years longer.

Not only did people die at a younger age a century and a half ago, but death was dealt with more openly and with a greater attention to ceremony. This was particularly so in Victorian England, where the queen herself, widowed at age forty-two, set the example for mourning. People generally did not die in hospitals or nursing homes at the time, but instead where they had lived their lives, in their own homes, with loved ones at their bedsides. I will never forget a line in H. V. Carter's diary. After receiving news of his mother's death, Carter asks himself, "What did I feel?

"Regret, mainly," he answers—regret that he hadn't been there.

THIS TIME, I had come to the anatomy lab by myself. There was no line of students waiting outside, and, in fact, the lab itself was almost empty when I arrived just before 8:00 A.M. The class I'd come to observe was called Epilogue. It was part of an intensive "refresher" course for second-year med students before they took their board exams. Frankly, I thought it might be refreshing for me, too, a chance to reconnect with some of my anatomy teachers and to get reinspired, a coda to my experiences at UCSF.

Though the class was set in the lab, the students would not be doing any dissecting that day. Prosections would suffice. To make room for the whole class, the dissection tables had been pushed off to one end of the lab and all the cadavers lay on top. Piled two to three to a table, they appeared to be huddled together for warmth, waiting, quietly waiting. Waiting to be of use. Waiting, it seemed at that moment, for me.

Remember how scary they'd seemed at first? I said to myself. The *thought* of what was inside those bags had been so much worse than the reality. I could still picture myself filing into the lab with the pharmacy students. No one spoke. It was as if there were forty pink elephants in the room—except that all forty were encased in bright white vinyl and bore the unmistakable profile of a human being: rounded skull, nose, mound at the stomach, jutting toes. I headed to the back counter, I recalled, and put down my things. As I eyed a body on a nearby table, I suddenly noticed its small feet poking out. I crept forward. Both feet were wrapped in gauze, which elicited in me feelings of sympathy and tenderness rather than fear or revulsion. *Oh, it's wounded,* I thought instinctively, illogically. With a closer look, I saw patches of mottled, brownish flesh on the shins, which didn't bother me at all. I unzipped the body bag all the way; I was ready to see more.

Later that same day, I started a diary. I had not kept one in well over twenty years and, unlike the diaries of my youth, this was meant to be just-the-facts. I simply wanted to get details of dissections and snatches of dialogue down on paper each night while they were fresh in my mind—an aide-mémoire for the writing of this book. It was not long, though, before I began including personal reflections. On October 10, 2004, for instance, I confided to myself:

> *I have to be honest. It's not just the body I'm fascinated by but also death. The snuffed-out, no-second-chances finality of it. The randomness of it. The nearness of it.*
>
> *Death has always seemed near to me. Even as a little kid, dying didn't seem eighty or a hundred years away—impossible to conceive. But instead, as if it could be close. Not that it would be, but could be.*
>
> *I still think about death all the time. I keep expecting it. Not my own, necessarily, but someone's. . . . I can feel it getting nearer and nearer, now and then even brushing up against me. Lying awake in bed, sometimes I feel it pressing against my body.*

Two years later to the day, at eight in the morning on October 10, 2006, Steve died in bed beside me. Though he was extremely fit and in excellent health overall, apparently a freak episode of cardiac arrhythmia led to respiratory arrest and, ultimately, cardiac arrest. I woke to the terrible sound and sight of Steve struggling desperately to breathe. Even more terrifying was the complete silence that soon followed, his body motionless. I started CPR, paramedics came, and we got him to the ER, but they were never able to get a heartbeat. Steve was forty-three years old.

WE HAD BEEN together for sixteen years. Steve was my partner not only in life but in writing, especially on this book—the Carter to my Gray, as I would affectionately say. ("No, make that the Gray to your Carter," he would tease in return.) After his death, going back and completing the final draft of the book seemed daunting. I wasn't sure if I could do it without him.

Though Steve never set foot in the anatomy lab, I even sensed his absence at Epilogue that morning. He used to drop me off before every class and pick me up after every lab and listen to my daily debriefings on the ride home. After the Epilogue class, I would have told him how I'd seen Dana and Kim and Dhillon and Charlie; how Sexton had retired, Anne had been promoted, and Andy had taken a new job; how all the med students were unfamiliar to me (Meri and Kolja and the others I'd studied with were already in their third year); and how, just as in the past, I was repeatedly mistaken by the students for a TA. When I told them I was writing a book about *Gray's Anatomy*, most students assumed I was talking about the TV show. "Yep, that's right," I'd say teasingly, "I'm telling the true story of Meredith Grey and Dr. McDreamy."

Finally, I would have told Steve that I'd been nervous to see the bodies, afraid that they might stir up upsetting memories of his death. By the way, I think that is one of the strangest things about losing a longtime partner: the very person you *most* want to talk to about your loss is the person who is gone. Well, it doesn't always stop me; sometimes I talk to Steve anyway.

"Honey-pie, you would've been proud of me," I told him as I drove myself home from UCSF. "Before class got started, I just took a deep breath and marched over to one of the tables and unzipped a body bag." The cadaver looked like a *cadaver*, I told him, and nothing at all like you. I laughed aloud, unsure how that had come out. "Well, you know what I mean," I said, and I was sure he did.

"It was like when I brought your ashes home," I continued. "To me, it was so clear that it wasn't *you* in that nice cedar box; it was just . . . your remains." He had long since left his body, I believed. I had seen it happen with my own eyes—life leaving him with his last breath.

You are dust, and to dust you shall return.

Officially, Steve was declared dead at the hospital. *Declared dead*—what a strange phrase. That makes it sound like an announcement was made over a loudspeaker. In fact, it was more like an unspoken exchange between the attending physician and myself. He left his position at the head of the gurney and approached me at the other end, where I was cradling Steve's feet. The doctor's pained expression told me everything I needed to know. I nodded, and practically in a whisper, he gave directions to his team. The medic, by now drenched in sweat, stopped CPR. A nurse shut off the respirator. And with that, everyone in the room quietly filed out and left me alone with Steve. Less than an hour after I'd awakened, I found myself performing a last rite, sacred in its intimacy: shutting Steve's eyes entirely closed with my fingers. I removed his rings, put them on, and said what he had not been able to say to me: Goodbye.

Nothing I'd learned in anatomy class prepared me for that moment. Nothing. Even being able to understand precisely what had happened to Steve anatomically and physiologically—I was easily able to read and interpret his autopsy report, for example—did not make it any easier to bear or grasp the fact that he was so suddenly gone. Which I see now as the painful final lesson of my education in anatomy. True, in a literal sense, I had never been nearer to death than in the lab itself. Over the course of a year, the result of a journey that began with Henry Gray and H. V. Carter and their book, I touched and felt and dissected dead bodies with my own hands and was constantly surrounded by dozens of them, to the point that I became inured to the sight. I gained a keen understanding of the *fabric* of the body—the raw, organic nature of flesh and bone and blood. But you don't learn about death from dead bodies. Just as you learn about the body by dissecting one, you learn about death by experiencing a death, by losing someone you love.

Conversely, what you learn about in Gross Anatomy is life, human life, clichéd as that may sound. In one of the last dissections I performed, I remember, I anatomized the knee, shoulder, and elbow joints, in effect dissecting the mechanics of human movement. And what is life—or, a defining sign of life—but movement? Whether the blinking of an eye or the wiggling of fingers or, with arms and legs pumping and lungs heaving, the running toward something in a great burst of speed—toward a goal, toward a finish line, flat out to the very end.

THE END.

Appendix

HENRY GRAY

1827–61

Gray's final resting place is London's Highgate Cemetery, where he shares the same grave as his mother, Ann, who died five years after him. *Gray's Anatomy*, now in its thirty-ninth edition in England and thirty-seventh in the United States, has never gone out of print and has sold an estimated five million copies to date. The year 2008 marks the 150th anniversary of *Gray's Anatomy*.

HENRY BARLOW CARTER

1803–68

The patriarch of the Carter family, artist Henry Barlow Carter, died of bronchitis on October 4, 1868, at age sixty-five. As described in his obituary, Henry Sr. sounds very much like his eldest son: "There was a natural reserve about him that rather prevented an extensive friendship, but those admitted into that circle were often charmed with the geniality of his spirit and the originality of his ideas."

JOSEPH NEWINGTON CARTER

1835–71

In 1859, H.V.'s free-spirited younger brother began teaching art and, along with sister Lily, converted to Christianity. "She and Joe walk as fellow Christians," a happy H.V. wrote at the time. In May 1871, Joe, a working artist, married longtime love Elisabeth Smith Newham, who was widowed three months later. Joe died of double pneumonia at their Scarborough home on August 16, 1871.

ELIZA HARRIET CARTER DI VILLALTA

1860–91

Henry and Harriet Carter's daughter, Eliza, married an Italian soldier, Federico di Villalta, in 1881. At the time of her death in Florence at age thirty-one (cause unknown), her only child, Ignazio Federico, was five years old. In his will, Carter established a sizable trust for his grandson.

Henry Vandyke Carter
1831–97

Carter's official cause of death was "phthisis pulmonalis" (pulmonary tuberculosis), and he is buried in a family plot at Scarborough Cemetery. Funds are currently being raised by the London County Council to commemorate his life with one of the city's famous blue plaques.

Eliza Sophia "Lily" Carter Moon
1832–98

Lily outlived her husband (William James Moon) and both brothers and died at her home in Scarborough on December 14, 1898, of "acute pneumonia." She was sixty-five years old. Lily was survived by three of her four children.

Acknowledgments

To the anatomists, teachers, fellow students, librarians, archivists, agent, editor, publisher, family members, and friends who helped me during various stages of writing this book, I thank you all equally:

Steve Barclay, Lea Beresford, Barbara and John Bourassa, Jane Breyer, Sylvia Brownrigg, the Byrne family, Gina Centrello, Andy Chamberlain, Susan Cohan, Ben Collins and Stephen Pelton, Doug Cooper, Nancy and Tim Cossette, Robin Coupar, Chris Davis, Josh Devore, Nripendra Dhillon, Tom DiRenzo, Anne Donjacour, Martin Duke, Meri Dunn, Amanda Engineer and the entire staff of the Poynter Reading Room at the Wellcome Library, Emily Forland, Sheila Geraghty, Sandra Gibson, Shawn Hassler, Jean and John Hayes, Patti Hayes, Mary and John Kamb, Lang Kehua, Rachel Kind, Yvonne Leach, Ming Ma, Massoud, Kathy and Dan Mayeda, Hazel McDonald, David Mikko, Nancy Miller, Keith Nicol, Charlie Ordahl, Ben Ospital, Chris Ospital, Kolja Paech, Sam Pak, Sandra Phillips, Kelly Piacente, Martin Pugh, Gregory Riley, Rev. Rachel Rivers, Richard Rodriguez, Dana Rohde, Conrad Sanchez, Cap Sparling, Kristen Stewart, Cheyenne Strube, Sexton Sutherland, Paul Teresi, Janine Terrano, Nallini Thevakarrunai, Kim Topp, Fernando Vescia, Jay Wagner, Wendy Weil, Vicki and Jim Weiland, Valerie Wheat, Paul Wisotzky, and Melaine Zimmerman.

Bibliography

Henry Vandyke Carter

Carter, Henry Vandyke. *On Leprosy and Elephantiasis.* London: G. E. Eyre & W. Spottiswoode, 1874.

————. *On Mycetoma, or the Fungus Disease of India.* London: J. & A. Churchill, 1874.

————. *Report on Leprosy and Leper Asylums in Norway.* London: G. E. Eyre & W. Spottiswoode, 1874.

————. *Spirillum Fever.* London: J. & A. Churchill, 1882.

————. Student's diary, 1853. Manuscript held in the Library of the Royal College of Surgeons of England, London.

The Carter Papers. Western Manuscripts Collection, MSS 5809–26. Wellcome Library for the History and Understanding of Medicine, London.

Gray, Henry. *Anatomy, Descriptive and Surgical.* Drawings by H. V. Carter. 1st ed. London: John W. Parker and Son, 1858.

Henry Gray

Gray, Henry. "An Account of a Dissection of an Ovarian Cyst." *Medico-Chirurgical Transactions.* Vol. 36, pp. 433–37. Published by the Royal Medical and Chirurgical Society. London: Longman et al., 1853.

————. *Anatomy, Descriptive and Surgical.* Drawings by H. V. Carter. 1st ed. London: John W. Parker and Son, 1858.

————. *Anatomy, Descriptive and Surgical.* Drawings by H. V. Carter, with additional drawings by Dr. Westmacott. 2nd ed. London: John W. Parker and Son, 1860.

————. *Gray's Anatomy.* Facsimile of rev. American ed. of 1901. New York: Gramercy Books, 1977.

————. "Injuries of the Neck." *A System of Surgery, Theoretical and Practical.* Vol. 2, pp. 270–339. Edited by Timothy Holmes. London: John W. Parker and Son, 1861.

————. "On Myeloid and Myelo-cystic Tumors of Bone, Their Structure, Pathology, and Mode of Diagnosis." *Medico-Chirurgical Transactions.* Vol. 39, pp. 121–49. Published by the Royal Medical and Chirurgical Society. London: Longman et al., 1856.

————. "On the Development of the Ductless Glands in the Chick." *Philosophical Transactions of the Royal Society of London* 142 (1852): 295–309.

————. "On the Development of the Retina and the Optic Nerve, and of the Membranous Labyrinth and Auditory Nerve." *Philosophical Transactions of the Royal Society of London* 140 (1850): 189–200.

————. *On the Structure and Use of the Spleen.* London: John W. Parker and Son, 1853.

GENERAL REFERENCE SOURCES

Bailey, Hamilton. *Notable Names in Medicine and Surgery.* London: H. K. Lewis & Co., 1959.

Bayliss, Anne M. "Henry Vandyke Carter." *Yorkshire History Quarterly* 4, no. 2 (Nov. 1998).

Bishop, W. J. "Henry Vandyke Carter." *Medical and Biological Illustration* 4, no. 1 (1954): 73–75.

Blomfield, J. *St. George's, 1733–1933.* London: The Medici Society, 1933.

Drake, Richard L., et al. *Gray's Anatomy for Students.* Philadelphia: Elsevier Churchill Livingstone, 2005.

Duke, Martin. "Henry Gray of Legendary Textbook Fame." *Connecticut Medicine* 57, no. 7 (July 1993): 471–74.

Encarta Encyclopedia. Standard ed. Microsoft: 2000.

Erisman, Fred. "The Critical Response to *Gray's Anatomy* (A Centennial Comment)." *Journal of Medical Education* 34, no. 1 (Jan. 1959): 589–91.

Evans, Alison, et al. *Human Anatomy: An Illustrated Laboratory Guide.* San Francisco: Regents of the University of California, 1982.

Goss, Charles Mayo. *A Brief Account of Henry Gray, F.R.S., and His "Anatomy, Descriptive and Surgical."* Philadelphia: Lea & Febiger, 1959.

———. "Henry Gray, F.R.S., F.R.C.S." *Anatomy of the Human Body by Henry Gray*. 29th American ed. Philadelphia: Lea & Febiger, 1973.

"Henry Gray." *St. George's Hospital Gazette* 16, no. 4 (May 21, 1908): 49–54.

Hiatt, Jonathan R., and Nathan Hiatt. "The Forgotten First Career of Doctor Henry Van Dyke Carter." *Journal of the American College of Surgeons* 181, no. 5 (Nov. 1995): 464–66.

Johnson, Edward C., et al. "The Origin and History of Embalming." In *Embalming: History, Theory, and Practice*, by Robert G. Mayer, pp. 23–40. Norwalk, Conn.: Appleton & Lange, 1990.

Moir, John. *Anatomical Education in a Scottish University, 1620*. Translated by R. K. French. Aberdeen, Scotland: Equipress, 1975.

Moore, Keith L., and Anne M. R. Agur. *Essential Clinical Anatomy*. 2nd ed. Philadelphia: Lippincott Williams & Wilkins, 2002.

Newman, Charles. *The Evolution of Medical Education in the Nineteenth Century*. London: Oxford University Press, 1957.

Nicol, Keith E. *Henry Gray of St. George's Hospital: A Chronology*. Privately printed by the author, 2002.

———. Interviews and correspondence with author, 2004–7.

Persaud, T.V.N. *Early History of Anatomy: From Antiquity to the Beginning of the Modern Era*. Springfield, Ill.: Charles C. Thomas Publisher, 1984.

———. *A History of Anatomy: The Post-Vesalian Era*. Springfield, Ill.: Charles C. Thomas Publisher, 1997.

Plarr, Victor Gustave. *Plarr's Lives of the Fellows of the Royal College of Surgeons of England*. London: Simpkin, Marshall, 1930.

Poynter, F.N.L. "*Gray's Anatomy*: The First Hundred Years." *British Medical Journal* 2 (Sept. 6, 1958): 610–11.

Roberts, Shirley. "Henry Gray and Henry Vandyke Carter: Creators of a Famous Textbook." *Journal of Medical Biography* 8 (Nov. 2000): 206–12.

Rohde, Dana. *Anatomy 116, Gross Anatomy: Course Lecture Syllabus*. San Francisco: Regents of the University of California, 2004.

Symmers, William St. Clair. "Henry Vandyke Carter on Mycetoma or the Fungus Disease of India." In *Curiosa: A Miscellany of Clinical and Pathological Experiences,* pp. 128–43. Baltimore: Williams & Wilkins Co., 1974.

Tansey, E. M. "A Brief History of *Gray's Anatomy.*" In *Gray's Anatomy.* 38th British ed. London: Churchill Livingstone, 1995.

Teresi, Paul, et al. *Prologue Block, IDS 101, Laboratory Guide.* San Francisco: Regents of the University of California, 2004.

Topp, Kimberly S. *PT 200, Neuromuscular Anatomy: Course Lecture Syllabus.* San Francisco: Regents of the University of California, 2004.

Williams, Peter L. "Historical Account: Biography of Henry Gray." In *Gray's Anatomy.* 38th British ed. London: Churchill Livingstone, 1995.

INDIVIDUAL CHAPTER SOURCES

CHAPTER ONE

Persaud, T.V.N. *Early History of Anatomy: From Antiquity to the Beginning of the Modern Era.* Springfield, Ill.: Charles C. Thomas Publisher, 1984.

Walsh, James J. *The Popes and Science: The History of the Papal Relations to Science During the Middle Ages and Down to Our Own Times.* New York: Fordham University Press, 1911.

CHAPTER TWO

Nuland, Sherwin B. *The Mysteries Within: A Surgeon Reflects on Medical Myths.* New York: Simon & Schuster, 2000.

Rohde, Dana. Interview with author. San Francisco, Mar. 23, 2005.

CHAPTER THREE

Hawkins, Charles. "London Teachers of Anatomy." *Lancet* (Sept. 27, 1884).

Holmes, Timothy. *Sir Benjamin Collins Brodie.* London: T. Fisher Unwin, 1898.

James, R. R. *School of Anatomy and Medicine, St. George's Hospital, 1830–1863.* Privately printed by the author, 1928.

CHAPTER FOUR

Griffenhagen, George B. *Tools of the Apothecary.* Washington, D.C.: American Pharmaceutical Association, 1957.

Nuland, Sherwin B. *The Mysteries Within: A Surgeon Reflects on Medical Myths.* New York: Simon & Schuster, 2000.

Trease, George Edward. *Pharmacy in History.* London: Baillière, Tindall, and Cox, 1964.

CHAPTER FIVE

Brodie, Benjamin Collins. *The Works of Sir Benjamin Collins Brodie.* Vol. 1. London: Longman et al., 1865.

CHAPTER SIX

Chadwick, Owen. *The Victorian Church.* New York: Oxford University Press, 1966.

Elliott-Binns, Leonard. *Religion in the Victorian Era.* London: Lutterworth Press, 1946.

"Religion in Victorian Britain" and "William Paley and Natural Theology." The Victorian Web. http://www.victorianweb.org/. Accessed Sept.–Oct. 2005.

CHAPTER NINE

"The Ant-Eater." *London Times,* Oct. 18, 1853, 9, col. 2.

Burke, Edmund. *A Philosophical Enquiry into the Sublime and the Beautiful.* London and New York: Penguin Books, 1998.

Hilton, Boyd. "The Role of Providence in Evangelical Social Thought." In *History, Society and the Churches: Essays in Honour of Owen Chad-*

wick, edited by Derek Beales and Geoffrey Best, pp. 215–33. Cambridge, U.K.: Cambridge University Press, 1985.

"Zoological Gardens, Regent's Park." *London Times,* Oct. 1, 1853, 8, col. 3.

CHAPTER ELEVEN

Arnold, Friedrich. *Icones Nervorum Capitis.* Heidelberg: Sumptibus auctoris, 1834.

Belt, Elmer. *Leonardo the Anatomist.* Lawrence: University of Kansas Press, 1955.

Calder, Ritchie. *Leonardo and the Age of the Eye.* New York: Simon & Schuster, 1970.

Choulant, Ludwig. *History and Bibliography of Anatomic Illustration.* Chicago: University of Chicago Press, 1920. Reprint of 1852 edition.

Nutton, Vivian. "Introduction." An online annotated translation of the 1543 and 1555 editions of Andreas Vesalius's *De humani corporis fabrica.* http://www.vesalius.northwestern.edu/. Accessed May 2006.

O'Malley, Charles D., and J. B. de C. M. Saunders. *Leonardo da Vinci on the Human Body.* New York: Henry Schuman, 1952.

Quain, Jones. *Elements of Anatomy.* 6th ed. Edited by William Sharpey. London: Walton and Maberly, 1856.

Richardson, Ruth. "A Historical Introduction to *Gray's Anatomy.*" In *Gray's Anatomy: The Anatomical Basis of Clinical Practice,* 39th ed., edited by Susan Standring, pp. xvii–xx. Edinburgh: Elsevier Churchill Livingstone, 2005.

Vesalius, Andreas. *De humani corporis fabrica (On the Fabric of the Body).* 2nd ed. Basil: Per Joannem Oporinum, 1555.

Vescia, Fernando. Interview with author. Palo Alto, Calif., Sept. 26, 2005.

CHAPTER TWELVE

Rohde, Dana. Interview with author. San Francisco, Mar. 23, 2005.

Shaffer, Kitt. "Teaching Anatomy in the Digital World." *New England Journal of Medicine* 351, no. 13 (Sept. 23, 2004): 1279–81.

Zarembo, Alan. "Cutting Out the Cadaver." *Los Angeles Times,* Feb. 28, 2004.

Zuger, Abigail. "Anatomy Lessons, a Vanishing Rite for Young Doctors." *New York Times,* Mar. 23, 2004.

CHAPTER THIRTEEN

Ordahl, Charlie. Interview with author. San Francisco, Oct. 4, 2004.

CHAPTER FOURTEEN

James, Lawrence. *The Rise and Fall of the British Empire.* New York: St. Martin's Griffin, 1997.

Ramanna, Mridula. *Western Medicine and Public Health in Colonial Bombay.* London: Sangram Books, 2002.

"Reviews and Notices of Books." *Lancet* 2 (Sept. 11, 1858): 282–83.

CHAPTER FIFTEEN

Carrell, Jennifer Lee. *The Speckled Monster: A Historical Tale of Battling Smallpox.* New York: Dutton, 2003.

"Death of Mr. Henry Gray, F.R.S." *Lancet* (June 15, 1861): 600.

CHAPTER SIXTEEN

"Death of Dr. Carter, of Scarborough, an Indian Medical Celebrity." *Scarborough Gazette and Weekly List of Visitors,* Thurs., May 6, 1897, 3, col. 3.

"Deaths: Henry Vandyke Carter." *Lancet* (May 15, 1897): 1381.

Gould, Tony. *A Disease Apart: Leprosy in the Modern World.* New York: St. Martin's Press, 2005.

"Honour to a Scarborough Gentleman." *Scarborough Gazette and Weekly List of Visitors,* Thurs., Dec. 18, 1890.

"Honour to a Scarborough Townsman." *Scarborough Gazette and Weekly List of Visitors,* Thurs., Nov. 20, 1890.

"Marriages." Henry Vandyke Carter to Mary Ellen Robison. *Scarborough Gazette and Weekly List of Visitors,* Thurs., Dec. 18, 1890.

"Marriages." Joseph Newington Carter to Elizabeth Smith Newham. *Scarborough Gazette and Weekly List of Visitors,* Thurs., May 4, 1871.

"Marriages." William James Moon to Eliza Sophia Carter. *Scarborough Gazette and Weekly List of Visitors,* Thurs., Feb. 12, 1863.

Pandya, Shubhada. "Nineteenth Century Indian Leper Census and the Doctors." *International Journal of Leprosy* 72, no. 3 (2004): 306–16.

Robertson, Jo. "Leprosy and the Elusive *M. leprae:* Colonial and Imperial Medical Exchanges in the Nineteenth Century." *História Ciências, Saúde—Manguinhos* 10, suppl. 1 (2003): 13–40.

APPENDIX

"Deaths." Henry Barlow Carter. *Scarborough Gazette and Weekly List of Visitors,* Thurs., Oct. 15, 1868, 3, col. 4.

"Deaths." Joseph Newington Carter. *Scarborough Gazette and Weekly List of Visitors,* Thurs., Aug. 17, 1871.

"Deaths." Eliza Sophia Moon. *Scarborough Gazette and Weekly List of Visitors,* Thurs., Dec. 22, 1898.

Index

Page numbers in *italics* refer to illustrations.

About the Author

BILL HAYES is the author of the national bestseller *Sleep Demons: An Insomniac's Memoir* and *Five Quarts: A Personal and Natural History of Blood*. His work has been published in *The New York Times Magazine* and *Details*, among other publications, and at Salon.com. He has also been featured on many NPR programs as well as the Discovery Health Channel. He lives in San Francisco. Visit the author's website at www.bill-hayes.com.

About the Type

This book was set in Monotype Dante, a typeface designed by Giovanni Mardersteig (1892–1977). Conceived as a private type for the Officina Bodoni in Verona, Italy, Dante was originally cut only for hand composition by Charles Malin, the famous Parisian punch cutter, between 1946 and 1952. Its first use was in an edition of Boccaccio's *Trattatello in laude di Dante* that appeared in 1954. The Monotype Corporation's version of Dante followed in 1957. Though modeled on the Aldine type used for Pietro Cardinal Bembo's treatise *De Aetna* in 1495, Dante is a thoroughly modern interpretation of that venerable face.